HF704175

# MUTATIONS

## Biology and Society

# MUTATIONS

## Biology and Society

Dwain N. Walcher, M.D., Director
*Institute for the Study of Human Development*
*College of Human Development*
*The Pennsylvania State University*
University Park, Pennsylvania

Norman Kretchmer, M.D., Ph.D., Director
*The National Institute of Child Health*
*and Human Development*
Bethesda, Maryland

Henry L. Barnett, M.D.
*Professor of Pediatrics*
*Albert Einstein College of Medicine of*
*Yeshiva University*
Bronx, New York

MASSON Publishing USA, Inc.
New York • Paris • Barcelona • Milan • Mexico City • Rio de Janeiro

The conference, MUTATIONS: BIOLOGY AND SOCIETY, organized by The International Organization for the Study of Human Development was made possible through the auspices of:

— l'Université René-Descartes,

— la Délégation Générale à la Recherche Scientifique et Technique,

— la Direction Générale des Relations Culturelles,

— la Délégation aux Relations Internationales (SEU),

—la Société Nestlé Alimentana,

— Ross Laboratories, Columbus, Ohio, USA

QH
460
.M87

Copyright © 1978, by Masson Publishing USA, Inc.

All rights reserved. No part of this book may be reproduced in any form, by photostat, microform, retrieval system, or any other means, without the prior written permission of the publisher.

ISBN 0-89352-020-9

*Library of Congress Catalog Card Number: 78-63411*

Printed in the United States of America

# Foreword

The word "mutation," from the Latin, mutare, which originally referred to all dramatic change, has in the past century been appropriated by biologists as a convenient term for designating heritable changes in the genetic material. Yet we know that in all social organisms, and particularly in man, mutation and social change are potentially synergistic during the course of evolution. For the student of human development, there is a need to understand "mutation" in its broadest sense—as a concept encompassing both genetic and social change.

Thus it seemed appropriate that the International Organization for the Study of Human Development (IOSHD)—concerned with an interdisciplinary approach to the complexities of human development—undertake this Conference on Mutation in conjunction with Professor Jean Frézal, the eminent physician and investigator in the area of human genetics, on the occasion of his retirement as President of the Université René Descartes.

The Conference planners sought to bring together authorities in the social and behavioral sciences and those from molecular and developmental biology for a scholarly discussion of the mechanisms and implications of change. We believe that the volume emanating from the Conference will be provocative, for this topic has never before been attempted by such a diverse group.

The volume begins with introductory remarks from Professor Ihsan Dogramaci, President of the International Pediatric Association, and Madame le Ministre Alice Saunier-Sèité, Secretary of State for Universities, France. These are followed by two essays, one from the perspective of a biologist and one from the viewpoint of a philosopher, placing the subsequent papers into context. The remainder of the volume consists of edited papers, some of which have been translated, that were presented at the Conference.

The volume is a tribute to Professor Jean Frézal, of the Department of Genetics, Université René Descartes, who has been and remains an internationally renowned figure in human genetics.

The editors are indebted to Mrs. Emily Walcher, to Mme. Agnes Cassiman, and to the National Institute of Child Health and Human Development, National Institutes of Health, U.S. Department of Health, Education, and Welfare, for help in preparing this document.

*Norman Kretchner*

# Preface

It is a pleasure for me to have an opportunity to say a few words on the occasion of the inaugural session of this International Conference for the Study of Human Development because I attach great importance to the themes it addresses. Mutations: Biology and Society; such is the theme wherein the two aspects are closely linked. We now see millions and millions of workers from developing countries working in industrialized countries. This occurrence, which often causes division of families and a harsh meeting of entirely different cultures, produces varied effects. The progress of technology and science poses equally new problems.

I am happy also that this conference has for its object a unique theme, that of mutations. We estimate now that limited meetings ought to be encouraged, and it is in this spirit that the International Pediatric Association (IPA), in recent years, has brought together five seminars on specialized subjects. Until then the IPA brought together, only every three years, the enormous International Congresses which attracted 6,000 to 7,000 participants.

I am happy to congratulate my friend, Professor Norman Kretchmer, for having chosen Paris as the site of the conference. Paris, in addition to its charm, holds a particular importance for the Association. It was, in effect, in Paris that the IPA was created 65 years ago and where the first Congress was held in 1912, under the leadership of Professor Henri Hutinel.

There are other reasons to believe that this congress will encounter great success. In the interest of the themes undertaken is added the scientific quality of the speakers, who are all very eminent specialists, among whom I cite Professor Jean Fourastié, member of the Institute, who will present an introductory report.

Finally, another factor of success lies in the fact that the congress will be presided over by Professor Jean Frézal. I know him not only as a scholar and eminent researcher, but I have worked with him during conferences of rectors, chancellors, and Presidents of European Universities, when, as President of René Descartes University, he represented all the French Universities. His statements in the course of these meetings and his profound knowledge of problems common to the University, on the European level, were particularly useful and appreciated by all the participants who were equally impressed with his profound courtesy. My admiration and my friendship date from this time. It is thus for me a great pleasure to see him preside at this conference, where, once again, I am sure he will assure its success.

*Ihsan Dogramaci*

You have selected mutations as the topic for your scientific discussions. You have stated your intention in the title of your Colloquium and, in a compact and rather provocative manner, have implied that you will try to reconcile the biological and social aspects of mutations; or, at least, that you will attempt to face this task without borrowing ideas from others but will state only your own.

Now, in the first place, what is a mutation? The French language dictionary defines it as follows: mutation is change. The examples which quote Littré indicate that it is an old French word which was originally used in this literal sense. Aubigné cited the example of the Queen of France "who would fear any mutation in the King;" that is, who would fear that he might change his opinion. Froissart also, in his chronicles, referred to "the sea which swells because of the mutation of winds."

This word has been used by naturalists such as Buffon, who spoke of "mutations of matter and form" as well as by historians and moralists. However, it is undoubtedly the latter who have had the greatest influence on its meaning and usage. When Bossuet, in his funeral oration for the Queen of England cried: "However high we may climb to seek examples of great mutations in the histories, we find that hitherto they were caused by laxity or by the violence of princes," he introduced a distinction between ordinary changes which affect the ordinary circumstances of life and the accident, the event whose unexpected and unprecedented consequences modify or upset the established order.

If historians are acknowledged to have a prior right to the word mutation, it must also be recognized that the wealth of meaning now attached to the word is the rightful property of a biologist, Hugo de Vries. It was de Vries, in his work on species and their varieties which was published at the beginning of the century, who advanced the concept that species are not gradually transformed by a series of imperceptible modifications, but by sudden and distinct changes. In the same way that peoples and societies undergo changes which upset them, nature also proceeds by leaps, and mutation is the means which she employs to produce new varieties and species.

Mutations, then, appear as a particular form of change, a change which brings about a rupture of continuity in a situation or in a state. In the tone of the word, we hear a note of crisis, suggesting a perilous and decisive moment or even a catastrophe, using this word in its most neutral sense of inversion or upheaval.

The most radical philosophies, particularly those that are associated with Marxist traditions, emphasize the inevitability of the transformation of societies, a process whose urgency derives from awareness of the incurable injustice of the present society.

Some schools of biological thought claim that every mutation in a species as admirably fashioned by evolution as our own constitutes a burden which contributes to the pollution and other nuisances produced by a proliferating and uncontrolled technology. At the extreme limit, this genetic load will threaten to become insupportable and will endanger the life of the species.

Although fundamentally different in its origins and inspiration, this fear can be seen to bear some resemblance to the political philosophy described earlier, in the sense that it sometimes leads to a radical disagreement with the technological society.

However, one may question the inexorable nature of the phenomena leading to mutations and also their predetermined character; that is, their tendency to obey laws of necessity. This observation leads to the thought that, despite the views of certain theoreticians, mutations could be and perhaps quite often are the consequence or expression of events which are intrinsically miniscule and indeed insignificant; they are, in any case, entirely fortuitous. To cite a few at random: the rise in price of a commodity, a technical innovation, or a change in the condition of a base in a molecule of nucleic acid.

This is not to say that one cannot identify phenomena which, by their individual importance or in combination with other events, act as the determining factors of an abrupt change, a mutation. Without doubt the demographic revolution, with the decrease in mortality and control of fecundity which accompanied it, constitutes the clearest paradigm of a major change arising out of the summation of minor events.

To state the matter simply, it appears to me that the complexity of the phenomena under consideration rules out any simplifying or global theories and requires the calm and objective study which you have had the merit and the courage to undertake.

One might continue this thought by considering not only the origins but also the ultimate effects of mutations, and I readily suggest that one would discern, from this point of view also, the complexity of the consequences and the fact that it would be simplistic to consider them as linear. In fact, the biological or social mechanisms which maintain or destroy equilibria must be better known before general conclusions can be drawn. It is in this manner that the question arises concerning the factors responsible for biological evolution and the role of chance; if this role is important, as is claimed by certain schools of thought, the thesis of genetic burden is demolished. At the same time, the idea that an abrupt change could be a precondition for the resolution of conflicts which arise from the fundamental relations between living beings, the sexes, age groups, or classes, and that this abrupt change could reestablish harmony between conflicting social agents, appears naive in many respects, if it is true that every society is problematic and vulnerable and that its future remains precarious.

*Alice Saunier-Sèitè*

# Contributors

**Barrai, Italo,** Professor of Genetics, University of Ferrara, Institute of Zoology and General Biology, Via Borsari 46, 44100 Ferrara, Italy

**Bernardi, Giorgio,** Directeur de Reserches au CNRS, Laboratoire de Génétique Moléculaire, Université de Paris VII—Tour 43, 2 Place Jussieus, 75005 Paris, France

**Boudon, Raymond,** Professeur, Université René Descartes, Groupe d'Étude des Méthodes de L'Analyse Sociologique, Maison des Sciences de l'Homme, 54, boulevard Raspail, 75006 Paris, France

**Cardo, Bernard,** Professeur de Psychophysiologie, Institut de Biologie Animale, Université de Bordeaux I, Avenue des Facultés, 33405 Talence Cédex, France

**Carlier, Michèle,** Maître Assistant, Clinique de Génétique Médicale, 149, Rue de Sevres, 75015 Paris, France

**Cavalli-Sforza, Luigi L.,** Professor of Genetics, Stanford University School of Medicine, Department of Genetics, Stanford, California 94305

**Destrade, M. Claude,** Maître en Psychologie, Laboratoire de Psychophysiologie, Institut de Biologie Animale, Université de Bordeaux I, Avenue des Facultés, 33405 Talence Cédex, France

**Dogramaci, Ihsan,** President of the International Pediatric Association, Chateau De Longchamp, Carrefour De Longchamp, Bois De Boulogne, 75016 Paris, France

**Dumont, Evelyne,** Chercheur D.G.R.S.T., Clinique de Génétique Médicale, 149, Rue de Sèvres, 75005 Paris, France

**Ebel, Alix,** Chargée de Recherche C.N.R.S., Centre de Neurochimie du C.N.R.S., 11, Rue Humann, 67085 Strasbourg Cédex, France

**Eckert, Roger,** Professor of Biology, Department of Biology, University of California, Los Angeles, California 90024

**Elster, Jon,** Associate Professor of History, University of Oslo, Postboks 1008 Blindern, Oslo 3, Norway

**Fourastié, Jean,** Professeur au Conservatoire National des Arts et Mètres (Paris), 10, Rue César Franck, 75015 Paris, France

**Girard, Alain,** Professeur, Department de Démographie, Université René Descartes (Paris V), 58, Rue dela Faisanderie, 75116 Paris, France

**Harris, Harry,** Harnwell Professor of Genetics, Richards Building/G4, School of Medicine, University of Pennsylvania, Philadelphia, Pennsylvania 19174

**Hirsch, Jerry,** Professor of Psychology and of Ecology, Ethology, and Evolution, Department of Psychology, University of Illinois, Champaign, Illinois 61820

**Jacquard, Albert M.,** Professeur et Directeur de Reserches, Institut National d'Études Démographiques, 27, Rue de Commandeur, 75675 Paris, Cédex 14, France

**Jaffard, M. Robert,** Maître-Assistant Université, Laboratoire de Psychophysiologie, Institut de Biologie Animale, Université de Bordeaux I, Avenue des Facultés, 33405 Talence Cédex, France

**Keyfitz, Nathan,** Andelot Professor of Sociology and Demography, Harvard University, 9 Bow Street, Cambridge, Massachusetts 02138

**Kretchmer, Norman,** Director, National Institute of Child Health and Human Development, The National Institutes of Health, Bethesda, Maryland 20014

**Larrouy, Georges,** Professeur, d'Hematologie, Université Paul-Sabatier, 37 Allees Jules Guesde, 31400 Toulouse Cédex, France

**Mattiuz, Pier L.,** Professor of Medical Genetics, University of Ferrara, Institute of Medical Genetics, Via Borsari 46, 44100 Ferrara, Italy

**Perruchet, Pierre,** Assistant, Clinique de Génétique Médicale, Université de Lille, I.R.O.P., 149 Rue de Sèvres, 75015 Paris, France

**Raveau, François,** Professeur et Directeur d'Études à l'École des Hautes Etudes en Sciencés, C.R.E.D.A., 45,Rue des Saint Péres, 75270 Paris, Cédex 06, France

**Ringertz, Nils R.,** Professor, Department of Medical Cell Genetics, Medical Nobel Institute, Karolinska Institute, S-104 01 Stockholm 60, Sweden

**Roubertoux, Pierre,** Maître Assistant, UER de Psychologie Clinique de Génétique Médicale, 149 Rue de Sèvres, 75015 Paris, France

**Royer, Pierre,** Professeur de Pédiatrie, Hôpital des Enfants Malades, 149, Rue des Sèvres, 75730 Paris, Cédex 15, France

**Ruffié, Jacques,** Professor, Laboratory of Physical Anthropology, College of France, Paris, France

**Saunier-Sèité, Alice,** Secrétaire d'État aux Universitiés, Paris, France

**Weissmann, Charles,** Professor, Institut fur Molekular-biologie I, Universität Zurich, 8093 Zurich, Switzerland

**Ziegler, Rolf,** Professor, Universität Wien, A-1080 Wien, Alsterstrasse 33, Austria

# Contents

# *Mutations, Nutrition, and Evolution*

Norman Kretchmer

The International Organization for the Study of Human Development was founded to bring together workers from a variety of disciplines concerned with the development of the human being. That decision implied that a number of people in this society would be speaking different scientific languages. Thus, the members would need not only to make an effort to understand one another, but also to speak in languages that make this possible. This conference is an example of this goal. Over the next few days we will discuss various cultural, biological, social, and psychological aspects of change--not all change, but change which is sudden, more or less permanent, and transmissible by hereditary. That is the change we call <u>mutation</u>.

I will have to speak as a biologist/clinician and use the language of those disciplines and I hope I will be understood even when I drift off into technicalities.

To the organism itself, the significance of a mutation depends upon the context in which that mutation occurs. We have already determined that a mutation is a significant and basic alteration and one that will not occur gradually but rather in a sudden and revolutionary fashion. Evenutally, the mutation can be expected to bring about a change in the phenotype of the organisms to which it is transmitted. If the mutation acts in a dominant fashion, the change may either be immediate or it may require several generations for full expression. In the case of a recessive mutation, the change may be either dramatic and apparent or it may be so subtle that sophisticated biochemical techniques are required for its detection. In either case, an individual with a new phenotype will emerge as a result of mutation; this individual will be subjected, along with the rest of the population, to selective pressure exerted by the environment. This pressure and the individual's genetically determined capacity to cope with it will determine the chances of the mutant gene being passed on to future generations. If the mutation proves advantageous to the individual in the present environment, the chances are that the mutant gene will be passed on and gradually increase in frequency in the population as a whole. If the mutation proves to be disadvantageous, it is more likely that it will be eliminated from the population, or retained at a very low frequency.

I am not going to enter into the difficult subject of neutral or very weak selected genes because there are other participants who are far better prepared to discuss this aspect of population genetics. The point I wish to emphasize is that the effect of many mutations depends on the environment to which the mutant individual is exposed--and here I mean social and cultural environment as well as the natural environment.

Let me give you an example. If we consider a pan-species trait such as the inability to synthesize ascorbic acid in the human, which must have arisen by mutation, we find that the effect of the mutation is only realized in relation to the nutritional circumstances of the individual. Studies of the skeletal remains of ancient hunting and gathering peoples yield little evidence of the sort of bone damage characteristic of scurvy; this finding suggests that ancient man obtained adequate vitamin C from the plant materials in his diet and there was no selective pressure against the mutation. However, the situation was quite different in the case of European seamen and inhabitants of European cities in the fifteenth to seventeenth centuries. The classic studies of James Lind indicate that scurvy developed in seamen who were denied access to fresh fruit for a long period. In this environmental circumstance the long-established mutation for inability to synthesize ascorbic acid suddenly became a lethal mutation. Later, as knowledge of nutrition increased, citrus fruits were included in the diet as a preventive measure, and deaths from scurvy became as great a rarity as they must have been in ancient times. This example illustrates the importance of environment, and particularly the part of the environment that we call nutrition, in determining whether a genetic trait will be advantageous, neutral, or disadvantageous. It also illustrates the manner in which cultural manipulation of the environment can bring about intentional or unintentional changes in selection pressure.

It is no longer necessary for us to dwell on past scientific arguments concerning the effect of nurture as opposed to that of nature. By now we are well aware that nature and nurture are inseparable, and we can discuss both in molecular terms that are finite, that have a particular language, and that have their code in DNA. We can examine particular molecular phenotypes and consider the mutational events and selection pressure from which they arose. We can also consider the environmental changes that would be advantageous to individuals possessing these various molecular phenotypes. Indeed, one of the major objectives of clinical genetics today is to find means for relaxing the selection pressure against individuals bearing deleterious mutations.

Food is one of the prime agents of evolution. The selective pressures exerted by nutrition affect every cell of every individual in the population. Nutritional science, in its broadest sense, will eventually provide highly significant insights into the pressures responsible for evolutionary change in higher organisms. If it is true that the genome of the organism and the gene pool of the population contain a cryptic history of past mutations and past selective pressures, we would expect that the message might be deciphered most easily by those with an understanding of nutrition.

In recent years there has been a great deal of research aimed at determining the exact effect of specific nutrients on specific molecular processes in higher organisms, with the goal of determining how particular nutrients exert selection pressure on a particular genotype. We have been able to show that the effects of certain deleterious mutations can be cancelled out by nutritional change. The best known example is, of course, the use of a diet low in phenylalanine to ameliorate mental retardation in children with phenylketonuria. Similarly, removal of lactose or galactose from the diet can prevent blindness in persons with another genetic disease, galactosemia.

Interest is now expanding beyond the relatively rare and deleterious mutations into the area of genetic polymorphism. We are beginning a systematic search for large-scale variations in nutritional requirements within and between human populations. By historical accident I became involved in the problem of nutrition as an agent of evolutionary change and have been fortunate enough to participate in a series of investigations that have contributed to the construction of a paradigm of the relationship between food and evolution.

In the intrauterine environment the mammalian fetus is exposed to simple and elementary foodstuffs that have been, in a sense, predigested by the mother--carbohydrates usually in the form of glucose and proteins in the form of their simplest component, amino acids. After birth, however, in all species of mammal and in all societies, the newborn animal is exposed to one food, milk, which is a complex mixture, made up of proteins, carbohydrates, and fat. Its synthesis represents a remarkable evolutionary response to the problem of securing a reliable, nutritious, and portable food supply for the newborn. The task of digesting milk and absorbing its constituents represents the first major nutritional challenge to the young mammal.

Milk is the product of at least 100 million years of evolution. The mammary gland is thought to have evolved from a sebaceous exocrine gland in an ancient Therapsid reptile. From this reptilian ancestor both the monotremes and the mammals are descended.

The proteins contained in milk provide clear markers of evolution and species variation. When the caseins in milk are examined carefully, as has been done by Ribadeau Dumas in Jouy, they can be shown to exhibit many complex polymorphisms that can be utilized as a basis for studying evolution and species change.

The carbohydrate, lactose, manufactured by the mammary gland, is unique. It is found in no other animal source, although a similar carbohydrate is found in some plants. Lactose is a disaccharide composed of two simple sugars, glucose and galactose. Its synthesis is accomplished by a curious mechanism involving two proteins, an enzyme generally called galactosyl transferase and a noncatalytic protein called alpha-lactalbumin, which acts as an enzyme specifier. By itself in many tissues, the enzyme catalyzes the synthesis of only N-acetyl galactosamine. However, in the mammary gland in the presence of alpha-lactalbumin, the enzyme catalyzes the production of lactose. This system is called lactose synthetase.

Alpha-lactalbumin has been studied in detail; its amino acid sequence indicates that this protein must have evolved from a much more ancient protein--the enzyme, lysozyme. Lysozyme is found throughout the mammalian body, particularly in tears, mucus, and macrophages where it serves to digest complex carbohydrates, particularly those of bacterial cell walls. Somehow, in the course of evolution, this digestive enzyme (or rather, a segment of DNA coding for it) gave rise to a nonenzymatic protein in the synthesis of lactose.

Lactose has several properties which might make it uniquely suited to its role as the chief carbohydrate of milk: it apparently cannot diffuse back into the bloodstream of the mother and it provides a most important carbohydrate, galactose, which is necessary for the synthesis of certain components of brain and nervous tissue. Also, it is a source of readily available energy for the newborn.

So far, only two exceptions are known to this

sequence of rather successful evolutionary changes. One group includes an oviparous anteater, echidna, which has the complex carbohydrate sialyl lactose (N-acetyl-4-O-acetylneuraminyl lactose) instead of lactose in its milk and the platypus which has difucosyl lactose. Since echidna and the platypus belong to the class Monotremata rather than the class Mammalia, this difference is not too surprising. However, the second exception is the really striking deviation from the typical mammalian pattern that occurs among certain sea mammals in the Pacific Basin. These pinnipeds, including the California sea lion, the walrus, and related species, have no lactose whatsoever in their milk. The energy requirements of the newborn are provided by fat which makes up 35 percent of the milk. Work done in our laboratory has shown that these Pacific Pinnipedia have galactosyl transferase in their mammary gland, but they have no alpha-lactalbumin. Consequently they are incapable of synthesizing lactose. We have not been able to determine whether lysozyme is present in the tissues of these animals; nor have we been able to determine what, if any, evolutionary advantage is conferred by the loss of the ability to synthesize lactose. Some authors have suggested that replacement of lactose with fat allows the animals to provide their young with twice as many calories per gram of milk. (Or perhaps the adult diet, which consists almost entirely of fish, has been so low in carbohydrate for so long that the animals have evolved unusually efficient mechanisms for the utilization of protein or fat as an energy source). Whatever the case with these exceptions, the fact remains that the ability to synthesize lactose must have proven advantageous for the vast majority of mammalian species.

Of course, inclusion of lactose in the milk presupposes some mechanism for the digestion, assimilation, and utilization of lactose in the newborn animal. The central feature of this mechanism is the enzyme, lactase, a disaccharidase present on the surface of the intestinal cells. Lactase specifically catalyzes the digestion of lactose and thus facilitates the assimilation of the components of lactose. The enzyme appears early in the life of the mammal, usually before birth, reaches its highest activity during the perinatal period, and essentially disappears at the end of suckling. In most humans the activity of lactase is considerably diminished by about three to five years of age.

If a young mammal with a congenital inability to synthesize lactase is fed milk containing lactose, the result will be a severe and possibly fatal diarrhea. This effect is observed in human infants suffering from congenital lactase deficiency, a very rare disorder apparently arising from a recessive mutation. Interestingly enough, it is also true of the Pacific sea mammals mentioned earlier. These animals, which have no lactose in their milk, have no lactase in their intestines at any age. If young Pacific sea lions or walruses are given cow's milk, they die. Thus, we have still another example of the manner in which specific nutrients can exert a selective pressure on individuals with specific genetic traits: When lactose is present in the diet in large quantitites, genetically determined lactase deficiency becomes a deleterious trait, and there will be a strong selection pressure against infants carrying that trait. trait. However, this selection pressure can be relaxed by removing the sugar, lactose, from the diet.

Unfortunately, very little is known about the sequence of steps involved in the evolution of the protein alpha-lactalbumin and the enzyme lactase. We cannot even guess which of them evolved first. It would be particularly interesting to know whether lactase also has an amino acid sequence related to that of lysozyme.

Another evolutionary problem which has intrigued many investigators is the phenomenon of lactose digestion in human adults. In most humans, lactase activity follows the typical mammalian pattern, reaching a peak in the perinatal period and declining to very low levels after infancy. In many humans, however, lactase activity remains high throughout life, allowing the individual to consume milk from domestic animals without ill effect.

Studies of the ethnic distribution of lactose digestion in adulthood have shown that this trait occurs with highest frequency among persons of Northern European ancestry, among the Fulani, Masai and Tussi tribes of Africa, and among persons descended from the Mongols of Central Asia.

Surely it cannot be a coincidence that these are the very peoples who have practiced dairying and have engaged in the consumption of fresh milk for thousands of years, and it cannot be a coincidence that persons who are pure descendants of aboriginal peoples living in the nondairying regions of the world--the Americas, Oceania, the Orient, and the coastal regions of sub-Saharan Africa--are universally incapable of digesting lactose in adulthood.

It appears that the domestication of animals and the initiation of the dairy industry, events thought to have occurred first in the Near East some 8,000 years ago, must have created a new selection pressure in favor of individuals having the genetically determined capacity for lactose digestion in adulthood. Family studies have shown that adult lactose digestion is inherited as a simple Mendelian dominant. The most logical explanation for the current ethnic distribution of adult lactose digestion, then, is that a dominant mutation arose among the ancient inhabitants of one of the centers of animal domestication, and that adults carrying the mutant gene enjoyed a selective advantage when lactose was introduced into the adult diet in the form of milk from domestic animals. The mutant gene would then have increased in frequency among those groups of humans who adopted milking as a cultural practice. The spread of the mutant gene would have been most strongly favored in groups who found it advantageous to drink their milk fresh rather than converting it into a fermented form. Nomadic or seminomadic cattle herders would certainly have benefited from the ability to digest this portable and highly accessible food. Thus, it is not surprising that their present descendants exhibit such a high incidence of adult lactose digestion.

In certain areas of the world historical evidence indicates that there has been recent admixture of ethnic groups which differ in their capacity to digest lactose; in such instances the mixed population descendants show an intermediate incidence of lactose digestion reflecting the contribution of each of the ancestral groups. Thus, about 30 percent of adult American Blacks are capable of digesting lactose--an incidence which correlates well with their estimated proportion of Northern European ancestry. The gene for adult lactose digestion has also entered the gene pool of American Indians and certain West African tribes (Yoruba, Ibo, Hausa) as a result of intermarriage with outsiders capable of digesting lactose.

So the current situation with respect to lactose digestion in adulthood is one of polymorphism. Whether the polymorphism will be transient or stable over the next century depends on the selection pressure exerted on humans possessing the original form of the gene vis-a-vis individuals possessing the mutant form. This in turn depends on the cultural pressure to drink milk. In countries populated largely by persons of Northern European ancestry, a belief in the nutritional virtues of milk has become so entrenched in the popular culture that the substance is urged on all citizens without regard to their individual abilities to digest lactose. The resulting clinical symptoms which so often accompany primary lactose malabsorption in the adult are considered to be pathological. However, what we are really dealing with here is a case of cultural error in which a particular food has been unwisely introduced into the diet of individuals and ethnic groups who are not genetically adapted to it. This food is now exerting a negative selection pressure against those individuals and those groups.

There is reason to suspect that milk is not the only food that is exerting an adverse effect on large numbers of humans. Dairying was only one of a number of cultural innovations which occurred in the millennia following the Ice Age. This period of innovation, which began about 10,000 years ago, has been called the Neolithic Revolution because it involved a radical change in the life-style of the peoples who passed through it. Human groups who had lived by hunting and gathering gradually domesticated plants and animals and became agriculturists or pastoralists. Their new food supplies permitted them to increase their population at a much more rapid rate, to congregate in cities, and to begin the process that we call civilization. However, their new foods radically altered their diet within a relatively short time and must have exposed them to a number of new selection pressures. Studies of adult lactose digestion and malabsorption have provided one well-publicized example of evolutionary change in humans arising out of cultural change occurring in Neolithic times.

Cultural history may also provide an explanation for striking ethnic differences in the incidence of another food-related disease, diabetes mellitus. It is now known that approximately 50 percent of the Indians of the American Southwest will develop diabetes at some time after 35 years of age. Neel has suggested that this type of diabetes, which clearly has a major genetic component, might be a reflection of a "thrifty genotype" which has evolved in response to the nutritional situation encountered by hunting and gathering peoples. Hunter-gatherers, according to Neel, might derive considerable advantage from genes for hyperinsulinism, as this would ensure that on occasions when they managed to gorge themselves with food, that food would be immediately placed in storage for future use. Unfortunately, hyperinsulinism in early life might also lead to obesity and eventual exhaustion of the pancreas in individuals continuously exposed to a high-carbohydrate diet, as is the case with the modern American Indian and the Eskimo.

Another food-related disease with a striking pattern of geographic distribution is celiac disease, or gluten-sensitive enteropathy. In celiac disease the protein, gluten, contained in wheat, oats, and barley, interacts with cells of the intestine in such a way as to cause malabsorption of nutrients, failure to thrive, and in some cases early death. Celiac disease occurs frequently in northern Italy, in the southern part of Switzerland, and in Ireland. In the Irish county of Galway, the frequency of celiac disease is 1 in 300, as compared with 1 in 3,000 in the United States. The incidence in the U.S. seems to be decreasing, either as a result of changes in wheat caused by selective breeding, or as a result of changing patterns of infant nutrition, or of genetic changes in wheat processing. Although celiac disease is probably inherited as a simple Mendelian trait, it does appear to have a genetic component that is somehow associated with the HLA-DW3 histocompatibility loci. Thus, celiac disease may turn out to be another example of genetically determined intolerance to a food that has entered the diet of certain populations in historic times.

In summary, we can say that if there is any message to be derived from the lactase/lactose story and the other examples cited in this discussion, it is that both mutation and ancestral culture have left their imprint on the genetic make-up of present-day human populations. Although mutation is a sudden organismal change, the sustenance of many mutations depends on the interplay between the biology of the host and the total environment, including cultural environment, over the course of generations.

## REFERENCES


1. Barnicot, N.A.: Evolutionary perspectives in human nutrition. In The Domestication and Exploitation of Plants and Animals Peter J. Ucko & G.W. Dimbleby (Eds.), Aldine-Atherton, Chicago and New York, 525-529, 1969.

2. Bokony, S.: Development of early stock rearing in the Near East. Nature 264, 19-23, 1976.

3. Brew, K.: Lactose synthetase: evolutionary origins, structure and control. Essays in Biochemistry, Vol. 6, 93-118, 1970.

4. Cavalli-Sforza, L.L., and Bodmer, W.F.: The Genetics of Human Populations., W.H. Freeman & Company, San Francisco, 1971.

5. Childs, B.: Sir Archibald Garrod's conception of chemical individuality: a modern application. New Eng J Med 282, 71-77, 1970.

6. Kretchmer, N.: Memorial lecture: lactose and lactose--a historical perspective. Gastroenterology, 61, 805-813, 1971.

7. Kretchmer, N., Rossi, E., and Sereni, F. (Eds.): Milk and Lactation. Modern Problems in Paediatrics, 15, S. Karger, AG, Basel, 1975.

8. Lewontin, R.C.: The problem of genetic diversity. In The Harvey Lectures, Series 70, Academic Press, New York, pp. 1-20, 1976.

9. Messer, M.: Identification of N-acetyl-4-0-acetylneuraminyl-lactose in echidna milk. Biochemical Journal 139, 415-420, 1974.

10. Scriver, C.R.: Realized and potential neutralization of mutant genes in man by nutritional selection.
Federated Proceedings 35, 2286-2290, 1976.

11. Simoons, F.J., Johnson, J.D., and Kretchmer, N.: Perspective on milk-drinking and malabsorption of lactose.
Pediatrics 59, 98-100, 1977.

# Evolution and Change

Jean Fourastié

I shall not discourse at great length on how highly honored and at the same time undeserving I feel to speak at this inaugural conference. I am indebted for this honor to the International Organization for the Study of Human Development and to the Organizing Committee of the present congress, and I thank them cordially. However, I feel unworthy when I consider the extreme difficulty of the problems to be addressed by the congress and the high competence and scientific culture of the specialists who have assembled for these discussions. The only valid reasons for my participation, then, are your indulgence and the fact that you do not expect anything from me other than a few thoughts and very general questions.

I shall not further belabor the multidisciplinary character of the work of the congress. Everyone knows that a multidisciplinary approach is at once desirable, fruitful, and difficult. Everyone in France knows that the recent reform of the universities has been quite disappointing in this respect, and that two colleagues from the same university can still ignore one another as long as they belong to two distinct faculties. It is, therefore, essential to establish concrete procedures for an exchange of viewpoints, for a confrontation of ideas, for results and methods. We must have meetings and round tables such as these, and Professor Frézal should be congratulated for having made a place for this in organization of the René Descartes University.

Furthermore, the scientific, social, political, moral, and philosophical importance of the problems

problems posed and approached here must be emphasized. The very name of the International Organization, founded in 1969 in Bethesda, Maryland, and presided over by Norman Kretchmer, bears witness to the seriousness of the issues under consideration. The titles of our various sessions and round tables confirm the fact. We are to consider wide variations in world demographic patterns; sociological theories of social behavior and perverse effects of decisions inspired by a concern for human welfare or arising from the aggregate impact of behaviors which are individually reasonable; nutrition; mechanisms and modifications of animal and human behaviors; mutations; and different types of polymorphisms and their links to certain diseases. All of these topics are at the same time subjects of meticulous research, in which only a very small number of persons can participate today, and also enormous problems whose solution or insolubility weighs heavily on the daily lives of our children and on the future humanity, so much so that the survival of the human species appears to be in doubt.

It is not only material problems such as biological, economic and social processes, or the political decisions of nations and of organizations which confront us, but also intellectual and spiritual problems, those that affect our zest for life, our happiness to exist, our faith in humanity. Today's human being, without kings and without gods, is expecting from you gentlemen, if not an immediate explanation of the universe and the meaning of life, at least some faint illumination of the process by which his superb progress of yesterday has today become a mockery, anguished or futile.

The theme of your colloquium is mutations. From that derives the title of this conference: Evolution and Change. The object of the following paragraphs will be to offer some reflections on the ideas of evolution, mutation, and change.

The most general fact which must be mentioned is that the sciences have revealed and continue to discover for us a universe, a world, societies, beings, and things--let us say "elements"--in a state of change, of modification, of transformation. Perceptible reality is not fixed, is not stable. Certain systems seem approximately stable according to a time scale of millions of years, but others appear, exist, and come to an end in a fraction of a second.

With reference to the scale of time which humans have come to understand, the word evolution is generally used for modifications which occur over a long term, change for modifications occurring over a moderate term, and mutations for modifications which occur over a very short term. It is evident that this terminology is rather vague, because three words could not possibly be adequate to specify amounts of time ranging from one trillionth of a second to 5 billion years or more. Each system of realities has its own time scale, and the terminology of time ought to be defined for each field of study.

The principal fact, and one which is common to the entirety of the known universe, is that the universe and most (if not all) of its elements are evolving, and that this evolution can readily be analyzed in terms of change and events of mutation.

However, the preceding paragraphs are phrased more in literary than in scientific language. In the language of science, only one of the three words has a precise meaning, and that only for biology: it is the word mutation. Outside of biology, none of the words discussed thus far (modification, evolution, change, and mutation) has a scientific meaning, although all of them are frequently used. It is for this reason that we propose the terminology given above, which will be more accurately specified later.

Let us remember that, in this terminology, the word modification is the most comprehensive term. It covers all shifts of identity, all transformations of an object or being, whatever they may be.

| | |
|---|---|
| | over a very long term = evolution |
| Modification | over a moderate term = change |
| | over a short term = mutation |

However, it must be noted at the outset that the relationships between evolution, change, and mutation, that is, between different temporal phases of the same long-term process, are exceedingly complex. There can be no doubt that evolution over a long term is constituted, engendered, oriented, and, as some believe, determined by moderate- and short-term modifications. However, it would be a serious error to deduce from this that sequence over a moderate term is always in the same direction as the sequence over a long term, or that short-term events follow the same direction as those occurring over a moderate term. This is not the case at all. For example, it may very well happen (although probably only in rare instances) that a particular evolution is the result of a very small number of mutations, or even a single one amidst a vast number of other mutations or changes which had no consequence at all or did not give rise to other evolutions.

It is for this reason that detection of actual evolutionary events always eludes experimental science and even the science of observation, because no scientifically competent observer can know *a priori* which modifications will be decisive in the long-term process of evolution, nor can he arrange to be present at the precise moment when these modifications occur. The science of observation can only determine different states at different times and assume that they represent links of the same chain, of which many or a few intermediate links are missing.

These facts make the study of all evolution, and likewise of all change, indeed even of all mutation (because within each time span there is an undetermined number of shorter intervals), a problem of many complexities, naturally open to debate and disagreement. At the extreme limit, it may be claimed that the great problems of evolution escape science altogether and are consequently in the domain of nonexperimental rationality alone, or perhaps in the realm of faith. All of the information contributed by science is nonetheless essential to their solution.

The preceding observations obviously fall short of settling or exhausting the question of the duration of modifications. It is necessary, even in this superficial survey, to address the problem of rates of change. These rates differ radically according to the systems, phenomena, objects, substances, and even phases in which they occur. It is therefore necessary to specify for each case, not only the appropriate time scale, but also the time required for the appearance of a modification in an element, and the period which must elapse before this modification can become generalized in a population comprised of an aggregate of these elements.

The study of modifications also implies a spatial scale, according to which change can be recorded in individual objects or beings and in aggregates of objects or populations of beings. In this context, it is necessary to recall the successive levels of organization in biology and society: atoms are organized into molecules, molecules into cells, cells into organs, organs into organisms, and organisms into societies. Organisms are composed of and modified by molecules; societies are composed of and modified by individuals. An individual modification of behavior of a living being, of a man for example, could have as its source a physiological analogue of a biological mutation (for example, a modification of the cervical information network), which is not necessarily genetic and even, in general, nongenetic.

Nevertheless, modifications of biological hardware, whether of genetic origin or not, often affect the software, behavior. Such biological-behavioral relationships are the essential object for discussion at the present Colloquium. These relationships are clearly present in many instances; yet they are always complex, often ambiguous and unstable, even sporadic, intermittent, and recessive.

Studies of biological-behavioral relationships introduce a number of new factors and ideas, not yet confirmed by a sufficient period of observation and experimentation, but manifestly essential today. These include effects of biological selection and of cultural regulation, cultural diversity, genetic polymorphism, histocompatibility, and description and modifications of the structure of the genome.

New procedures, methods, and intellectual tools are undoubtedly necessary for the discovery, description, and comprehension by the human mind of these complex realities.

Now we must attempt to specify the various types of complexity that are suggested by these phenomena. We have already suggested a classification of modifications according to their relative duration, so that phenomena occurring over a long term are designated as evolutions, those occurring over a moderate term are called changes, and short-term events are referred to as mutations. It now seems useful to introduce a further distinction between 1) modifications which arise from simple displacements of the constituent elements of a system, without alteration of the relationships between these elements, and therefore result only in a spatial transformation of the system; and 2) modifications which affect the relationships between elements, and consequently give rise to a transformation in the nature of the system. An astronomical movement, such as the motion of planets around the sun, is an example of the first type; a genetic mutation is of the second type.

Perhaps it would be of interest to distinguish between environments (aggregations of elements) in which phenomena of the first type generally occur and those in which phenomena of the second type generally occur. Thus we can envision a distinction between systems, aggregates, substances, organs, organisms, and societies.

We can now refer to the modern discipline of systems analysis and the principle of treating systems differently according to their essential properties.

Classical and modern principles of logic, including Cartesian rationality (that is, principles elucidated by Descartes or their recent derivatives) have performed unceasing wonders in the discovery and description of what we have just called "aggregates." However, proof is accumulating as to the inadequacy or mediocrity of this logic (even though it is becoming more and more specific and complicated) when applied to the discovery and description of substances, organisms, or societies. This inadequacy is the basis for the contemporary invention of systems analysis or systems theory, a discipline which is still in its infancy.

One of the pioneers of systems theory, Pierre Vendryès, has just published an article in the _Bulltein biologique de la France et de la Belgique_ that deserves a great deal of attention (1). Vendryès suggests that modern biology will prove to be much more compatible with the logic of Aristotle than with that of Descartes. The logic of Descartes, which is the rationality currently in general use, recognizes only one kind of cause, the determinant of effects; but the logic of Aristotle acknowledges four types of causes: material, motive, formal, and final. Elaborating on his previous analysis of autonomous systems, Vendryès envisions a coherent theory of autogenesis, encompassing the generation and degeneration of substances.

Without going into the details of these difficult and controversial problems, I should like to state that I am personally impressed with the tendency of the atom to enter into a state of organization. A given atom, particularly a hydrogen or carbon atom, is likely to enter into an indefinite or practically infinite number of combinations, which are, nevertheless, far from random. Of these combinations, some are aggregates or collectivities and others are substances. Among the substances, some, such as proteins, are formed in a specific manner (or informed, in the sense of receiving information); others have a specific property of giving rise to form (or informing), as in the case of DNA.

The questions raised in this discussion are only those of an ignorant person of good will.

They represent only a small portion of the difficulties and mysteries confronting the participants in this Colloquium. It is on the basis of reality, that is to say, on the basis of your discoveries, that our explanation of the world and of the human condition will be constructed.

## REFERENCES

1. Pierre Vendryès, "L'autogenèse de la cellule et la théorie d'évolution," in Bulletin biologique de la France et de la Belgique, 1978, pp. 253-282. Pierre Vendryès has expressed his "theory of autonomous systems" in his book Vie et probabilité (Life and Probability), Albin Michel, 1942. More recently, he has clarified and developed his reasoning in Vers une theorié de l'homme (Toward a Theory of Man), P.U.F., 1975.

2. For the application of systems theory to modern economic, social and political "societies," Les systèmes du destin (Systems of Destiny), by Jacques Lesourne, Dalloz, October, 1976, is suggested as a reference.

3. A classic on the theory of systems is: General Systems Theory as a New Destiny, by W. Ashby, 1958.

# New Demographic Regime and the Control of Fertility

Alain Girard

Human fertility* has never been entirely natural. Customs of all sorts, moral or legal codes, both written and unwritten, and various social and cultural restrictions on sexual relations have combined to hold the population at a level below its reproductive potential. Except for a few groups and/or a few populations at specific times in their histories, until the dawn of the last two centuries the pattern has been for married couples to yield to the laws of chance and welcome what children were born to them as the means of perpetuating their kind.

Today, however, the industrial countries have espoused a radically different system. Following a transitory, semispontaneous, semivoluntary phase, reproduction is more and more controlled and directed. The invention of very safe contraceptives and birth control procedures, including legalized abortion practiced under adequate conditions of safety, allows us to believe that soon there will be no more births without the express decision of procreators. Reproduction ceases to be a chance phenomenon. People regulate it according to their wishes. That which has previously been submitted to or accepted now occurs only voluntarily. Procreation is becoming a matter of belief and/or the result of a calculated opportunity.

---

*Throughout this discussion, fertility will be used to refer to actual production of offspring, while fecundity will be used to denote potential to reproduce.

Thus, by means of knowledge and acquired techniques, man can now avoid conditions whereby his own reproduction would be effected. A question arises: Will control of fertility place in jeopardy certain basic elements of the social equilibrium and alter, for the human species, what Jacob has called la logique du vivant (8).

## DECLINE OF MORTALITY AND GROWTH OF POPULATION

At the beginning of this era, and in consequence of scientific progress and subsequent alterations in living conditions, a new phenomenon appeared in the history of humanity: man's victory over many of the causes of death. This was a relatively recent phenomenon which appeared first about the middle of the 18th century and was confined at the outset to western and northern Europe. Since then it has spread over greater areas, has affected ever-increasing numbers of people, and has mitigated or conquered more and more mortal conditions. This great revolution has become both a symbol and a summary of all revolutions.

Very recent progress in the battle against prematurity and perinatal mortality, still unexpected a few years ago, enables us to project that every fetus, sufficiently advanced in growth and not aborted voluntarily, can survive birth and is increasingly likely to attain a life span equal to the rest of the population. Even if one discounts artificial insemination, already practiced voluntarily by some couples, it can be said that the success obtained against sterility assures almost an equal chance of procreation to all those desiring it. The human condition has not changed: each representative of the species remains mortal, but the process of life has acquired a new aspect.

The first result, and the most apparent, of the decrease in mortality and the increase in the mean duration of life has been the growth of population, first in Europe, then in the countries inhabited by peoples of European origin, and most recently in the Third World. This increase in population has been more striking and has occurred at a more rapid rate than the decline in mortality A number of sociologists, including Comte, Coste, Durkheim in particular, and Dupréel (3), who eventually changed his opinion, have considered technical progress to be a somewhat mechanical side-effect of population growth. Thus we arrive at the old question: In order to produce wealth, must we first have an abundance of people? Or must we first distribute wealth, in order for the population to grow? This question, as is well known, has acquired a new dimension today.

Actually, the effect of the two variables, technology and demography, has been concomitant in the industrial countries, the progress in one accompanying progress in the other. Thus, the evils predicted by Malthus have not materialized. However, there remains the threat that billions of people added to billions may foreshadow some as yet unknown disaster. There has been an appeal for slowing the growth of population, if not arresting its growth altogether, and many have extolled the benefits of maintaining a stationary state. Whatever the advantages or inconveniences of zero population growth, and well before it is realized, population imbalance may occur on a global scale, as it has already occurred on the scale of continents or regions. Apart from the tensions or conflicts which could result from such an imbalance, the number of people in itself carries a specific weight. Will a change in the dimensions of human societies lead to a change in their nature? What effect does the multiplication of people have upon the dispersion of genes, as well as upon social relations, within each country and at the international level?

## EXPLOSION OF SINGLE PERSONS AND DIMINUTION OF ENDOGAMY

The expansion of population volume is accompanied by another well-known phenomenon. The distribution on the earth has changed. Industry first and services next have drawn their workmen from the rural areas where a surplus of population had accumulated because of new agricultural techniques. Thus, instead of a sparse humanity, much more equally distributed over the whole territory and struggling for a place for survival, there was substituted a much more concentrated population in vast vertical groups. The rural districts were, at the same time, urbanized and today benefit from the same advantages as do the cities. From the viewpoint of reproduction, there emerged a mixed population. The choice of mate was no longer limited to those found within walking distance, in one's own village, or in adjacent villages. An unexpected result was that the coefficient of consanguinity, as Sutter (15) has shown, rose in the rural districts during the transitional phase of the demographic revolution. In effect, with the drop in mortality rate, the number of survivors per family tended to increase, producing a larger number of consanguineous unions and with them a larger incidence of hereditary disease. It was not until the beginning of the 20th century that there came a fragmentation of isolates and a decline in endogamy.

An urban population is not an undifferentiated mass. The non-natives form groups by provincial origin or ethnic characteristics. Quite a strong segregation also occurs in accordance with occupational activities and by income levels. These groupings constitute true social isolates, but their size far exceeds that of the isolates of past centuries. Even though proximity of geographic or social origin, together with similarity of physical or intellectual characteristics, continue to be dominant factors in the selection of a mate, the possibilities of meeting prospective mates are nevertheless enlarged, and endogamy, with all its attendant risks, has diminished.

On the other hand, it may be said that with the survival of virtually all neonates guaranteed by medical science, individuals who would formerly have disappeared through natural selection now attain adulthood and reproduce, perpetuating genetic traits which are pathological or even lethal.

## VOLUNTARY LIMITATION OF BIRTHS AND REDUCTION OF FERTILITY

The transition to the new regimen of fertility originated with the reduction of mortality. In a totally unconscious manner, and no doubt by a necessary process of adaptation, couples have commenced to limit their progeny voluntarily. Understanding that there is no longer a need for five or six children to assure the renewal of families or groups, they have learned how not to conceive as many children as nature could give them. Furthermore, they have not depended on the discovery of modern methods, chemical or mechanical, for even before these methods were invented, the net amount of reproduction in a number of countries was often found to be below unity, a sign that the generations were no longer replacing themselves in terms of number per number.

Today, birth control information and new procedures permit persons to plan to have as many children as desired, at intervals of their own choice. The average size of the family has reached a limit, as well as the interval at which mothers bring children into the world. The possibility of bearing children with congenital malformation is more likely among very young mothers or relatively old mothers. An example is found in mongolism (Down's syndrome), whose incidence increases with the mother's age. Such circumstances could have repercussions which might endanger the renewal of generations and could be accompanied by genetic impoverishment.

As is well known, a reversal occurred immediately after the war and continued in the years that followed: fecundity increased, resulting in resumption of population expansion at a reasonable rhythm. However, events move faster today. This variable could not have been other than an accident, a transitory variation of the secular trend, which did not imply a profound change of attitude. In the past ten years, in fact, fertility has resumed its downward movement in all the industrial countries, dropping to a level that had never been reached previously, showing that the number of deaths exceeds the number of births in many countries, a pattern which will present itself in other countries soon if nothing intervenes to change this course.

This collective decrease appears to correspond to individual intentions. In fact, as very many surveys have demonstrated (6), the ideal family size is seen to decrease regularly in all industrial countries with the pattern being two or three children, the value of two tending to prevail more and more over the value of three. Now, it is not inconceivable that with an ideal of two, the effective drop could attain a lower figure, although under present conditions of mortality and taking into account the sex ratio of births, it is now 2.1 children per woman, which assures replacement of generations. Once again, without pronouncements regarding the advantages or inconveniences of the status quo, or even a slight regression, such a situation poses a question regarding the future of certain communities, although others may grow by two or even three percent per annum. Counting its population and computing the future, Europe may again become subject to the anguish of a void.

A threat arises, which has more of a social than a genetic character, and which could be one of the paradoxes, if not the contradictions of tomorrow. Whereas previously a woman's whole adult life had been occupied with gestation and the nursing and care of the five or six children necessary for perpetuation of the family, today she is liberated from many maternal and domestic chores. In agreement with her partner, she now bears him two children at a selected time. Thus she finds herself available for other tasks, if she so desires; she has freedom to pursue her own interests. The statistic of 2.1 children may mean nothing to the individual, but the State, considering statistical computation, may wish to intervene. If the State is threatened by overpopulation, the State could raise the marriageable age, separate the couples, propose or even impose sterilization. On the other hand, if the political choice is for growth, the State could modify the legislation for abortion, or in a less official manner, limit the number of interventions which have been decided upon in certain instances. Thus the freedom given to an individual or couple through science and the new living conditions could be withdrawn in the name of community demands.

However, let us forget the threat and return to the family model. In proportion to the decrease in size of the family, as well as to its actual behavior, the differences in attitude which once separated the various social groups are becoming progressively attentuated. The closer a minimum threshold is approached, the more widespread the uniformity becomes. The variation between groups becomes blurred at the same time as variations within groups. The control of fertility diminishes the uncertain differences: the more fecund couple may limit its progeny to the same level as another which is less fecund.

Many very large families have in the past practically assured, on their own, the growth of the population. This is becoming an anachronistic phenomenon. The variance decreases simultaneously with the mean. Thus, in any case, if one dares say it, the aptitude for reproduction, the quality of each individual, and the probability of transmission of all the problems, tend to become equalized.

## THE AGING OF THE POPULATION AND THE AGED

The interplay between the two associated variables, mortality and fertility, may alter the course of human life in yet another way. Prolongation of life, coupled with an increase in life expectancy, inevitably increases the absolute number of aged or very aged persons in a given population, and the decline in reproduction guarantees an eventual increase in the proportion of older persons, so that gradually the young are replaced by the old. Thus, before the era of contraceptives, those over 60 years of age in the industrial countries accounted for 5 to 6 percent of the population. Today they comprise 17 to 18 percent. In a stabilized population with a low mortality rate, where the births and deaths are in equilibrium, the number of 50 year old persons will be about the same as the number of 5 year old children. Those over 60 would comprise about 30 percent of the population. We have not yet reached this point, but the population of the developed countries is becoming older. Right now three generations, of which two are adults, live simultaneously, as if superimposed upon one another.

Certainly the dilemma posed by Sauvy--to grow or to age--is arithmetically true. A population which increases by the arrival of more numerous generations will be rejuvenated; a population which does not increase in this fashion will grow older. Unlike the biological process of senescence, the aging of a population is not irreversible. Aging is not necessarily an evil in itself, if progress in productivity assures a better life than heretofore for children, adolescents, and retired people who do not participate in active work. It is, in fact, something to have a clear conscience about and from which many consequences could develop.

Mead has expressed the idea that parents of today are like immigrants in the world where the young live, because of the speed at which cultural changes have occurred (10). Perhaps as a result of the new redistribution of ages, and because their role in production is no longer necessary, the young find themselves in a state of increased dependence upon their elders. The period in life we call adolescence--the intermediate stage between childhood and adulthood--is now recognized, though in earlier periods it did not seem to exist. It is a time when individuals who have not yet been absorbed into occupational activity are exempt from responsibilities. Corresponding to a more precocious physiological puberty, favored by living conditions, there is now a sort of prolonged psychological puberty. The means for responding to the aspiration of the young to be introduced into society has not as yet been found. The different ages of life have changed. Their respective status must be redefined.

Thus, limitation of births carries with it a variety of consequences, because it represents a departure from the conditions which originally governed human reproduction. If man is less subject to nature and to chance, if he follows his own preferences, he may find that the wishes of individuals no longer correspond to collective expectations, or even, if there is a correspondence, many years may elapse without its being realized in practice. The economic situation, apprehensions for the future, or simply the style of life could induce couples to limit their progeny at times, and at other times encourage them to have one or more additional children. The repercussions of such decisions can be seen in the peaks and valleys which have characterized the birth rates for almost a century, and in the concomitant irregularities in the age profile of the population. A number of examples can be cited, and nothing is more damaging to society than spurts of growth succeeding reductions in births, or the reverse. In any of these cases there could be a temptation for governmental powers to attempt to intervene.

And what will happen if tomorrow a prospect which is already in view, the choice of an infant's sex, is realized? Some more or less serious distortions may threaten the distribution of the sexes, and will in any case jeopardize the harmonious renewal of the population.

## DEPRIVATION, UNIFORMITY

Growth of the population, urban concentration, and the modification of age gradation have had other profound repercussions. It is rather curious that today biology demonstrates that there are no two humans alike among all the multitudes because, except in the case of identical twins, genetic inheritance of each individual is different (2). There had been little doubt of this in past centuries, as evidenced by Pascal and Leibnitz, who defied their contemporaries to find two bunches of grapes or two leaves of a tree which were identical. Has not each one of us always regarded himself and been regarded by others as unique and irreplaceable? However, biology today draws new consequences from its rigorous knowledge.

The postponement of death has presented to each human being, now capable of blossoming like a plant tenderly cultivated, a value of his own, a specificity which he never before possessed. Individualism was born out of the demographic revolution. Life is no longer a matter of chance, a free gift which we accept without question, all the while resigning ourselves to the disappearance of those who do not survive. Death is no longer the necessary complement to life. Today death appears as an offense, and it is not surprising that so many books are being published on the subject.

We can refer here to the analysis offered by Fourastié in a 1959 article on the life span of the average man (5). These observations have lost none of their relevance in recent years. Fourastie notes that death has practically disappeared from the external landscape. It is no longer a fortuitous event that could strike at any moment, particularly among the elderly. Death is now repudiated. The peasant of old accepted all the miseries and catastrophes that fate dealt him, and life resumed its course after epidemics or famines. However, the urbanite of today has before him the whole span of life. Diseases and physical suffering have been ameliorated; their place has been taken by psychological distress, depressions, and neuroses. Thus pain has not disappeared; it has merely changed its form.

The social pathology of the past, during the transitional phase, undoubtedly had its roots in the profound divisions which arose between various social groups as they became increasingly differentiated from one another. There appeared a pattern of mortality strongly determined by social ranking, a true "inequality in the face of death," accompanied by a no less sharply defined gradient in fertility. The less favored groups continued to procreate and die relatively early, while the privileged learned to calculate their personal chances for the future. The ideology of class war appears in historical perspective as an expression of this divergence of attitudes which arose in the course of demographic evolution of different social groups.

Today the range of diversity is narrowing; the trend is toward uniformity. Regardless of social or economic status, people come closer and closer to having the same dwellings or automobiles, the same number of children, and the same life expectancy. According to Laslett, we are becoming a single class society (9). We are leaning more and more toward a uniform state. The symptoms of future pathology are already inscribed in the present society.

A paradox is apparent: whereas the number of people has not ceased to grow and people live in more and more populated communities, the conscience of the group appears to be in the process of decomposition, as though reduced to atoms. The sociologists as they have measured the road that has been traversed have been saying this since the end of the 19th century. They place a new world in opposition to an old world, substituting society for community according to Tonnies, organic solidarity for mechanical solidarity according to Durkheim, and secondary relationships for primary relationships according to Cooley. Though the language differs, the reality remains the same. The individual in the anonymous crowds of the cities keeps close to the others at every moment, but he maintains only functional relationships with them and perceives only impersonal faces, regarding them with a feeling of anguished loneliness.

## THE FAMILY

The individual's refuge is in a life of privacy, in the heart of the family. Children--historically, first the young boy and more recently the young girl, whose sole function is no longer to become a mother--have occupied an increasingly larger place in the family; so much so that, as Philippe Aries has observed, it is now possible to distinguish "two types of family, before the child and after" (1). According to Louis Roussel (14), the trend has proceeded so far that the roles of parent and child are becoming reversed. Adults are turning to their children for the warmth and personal communication which can no longer be found outside the home. Parents are now the demanders of attention. These anxious parental demands cannot be directed to a large number of children, and only a very small number are actually required to satisfy them. This trend toward parental dependence may explain the accompanying decline in the birth rate, since all things are connected. It also imposes on the child a responsibility very difficult to bear.

Another significant observation is that the family has not been reduced to a nucleus consisting of a couple and those of their children who are still young. There is good reason to revise Parsons' theory on this point (11). The life span is lengthened and the children, when they achieve autonomy, do not disappear from the familial horizon. Instead there exists between the generations "a solidarity which is manifestly very close," in the words of Roussel. To be sure there is no longer cohabitation by parents and their married children, but there is a high degree of proximity of residences and frequent exchange of services. Certainly, parents now hesitate to dictate to their children the values upon which they founded their families; they allow young adults the freedom to make their own decisions, but continue to require their affection though without actually voicing the demand too often. From the oldest to the youngest in the family an unstable equilibrium prevails as a "compensation for the diminution of other solidarities."

This view of familial relationships must be considered in conjunction with certain questions arising from examination of recent demographic indices. Something is shifting in the very foundations of the family: there has been a drop in marriages and a corresponding increase in cohabitation without marriage, a rise in illegitimate births, and a perceptible increase in the divorce rate. These events, coinciding for the past ten years with a decrease in reproduction, are more pronounced in some industrial countries than in others, but it appears that we may well be experiencing a universal trend. Once again we cannot predict the extent of these changes; nor can we be certain whether the cohabitation of young people represents a new form of preparation for marriage or the prelude to a more radical overthrow of legal ties between individuals.

## THE RIGHT TO LIFE AND PROTECTION

From the moment when the security and aspirations of the individual begin to assume in his system of values a greater importance than the security and aspirations of the group, and when the well-being of the individual becomes an imperative responsibility for society, the traditional institutions which have assured social stability will necessarily be unsettled. There follows a crisis of values and a constant questioning of societal goals. The right to life, given to all, is accompanied by a whole series of other rights for each individual: the rights to education, to work, to leisure, and to many other satisfactions. A claim to equality seems to be included in the right to life.

But all of these rights, and this equality, the individual is empowered to obtain by himself. He demands the security which was not claimed yesterday by others, he demands to be protected, and he expects this protection to come from the outside. His requirement of external security is a symptom of his internal insecurity. To material insecurity has been added psychological insecurity. Eventually the individual, forgetting his uniqueness, turns toward others and tries to unite with them in order to define a common objective, utilizing unions or other organizations which can speak for him and assume the responsibility of insuring him fulfillment of his rights and satisfaction of his needs--even perhaps his need not to be left alone. The right to protection is becoming the most important of all the rights. Thus, lost in the "lonely crowd," one passes from an "inner-directed" to an "other-directed" character, to employ Riesman's terms (12).

Emphasis is no longer placed upon the essential preoccupation of individuals living during the transitional phase--the need to find within oneself the power to act, and to observe and evaluate one's own progress day after day. Models of behavior are now borrowed from outside, from the surrounding environment, across a sort of generalized conformity, in order that many may escape from his dissimilarity to others. Isn't there a dangerous aspect to this current tendency to demand independence while rejecting all interpersonal differences, and to demand freedom and protection simultaneously?

None of the periods covered by the demographic revolution is entirely separate from the others. Something of the old state exists in the new state. The trends are not so engrained that it would be hopeless to oppose them. In fact, despite the uniformity and growing homogeneity which seem to derive from the control of reproduction, one aspect of demographic change is certainly left to chance: the large-scale migratory movements which invariably violate any limitations adopted by the restrictive groups in which individuals now live. This element of danger undoubtedly arouses our interest in protection.

Once again, because of all the progress which has been accomplished, of which the change in the demographic area is only an expression, but which rests upon the carnal subsoil of our existence, we find ourselves in the presence of an unknown future which we are incapable of imagining. The human race, wrote Valéry with astonishing foresight of the circumstances which are confirmed today, "is perhaps far removed from the primitive and natural conditions of the species," having arrived at a phase in its history "where every forecast - only because it is a forecast - becomes a chance of error, a suspect production of our fancy" (16). The demographic perspectives, if often contradicted by facts, actually prompt us to caution.

A "sudden mutation," again an expression by Valéry, is nonetheless accomplished. Its consequences cannot be foreseen; we will undoubtedly find it appropriate to rely upon the faculties of adaptation which man has always manifested. The new era of human reproduction is nothing other than an adaptation to a new era of mortality. However, the destruction of an old equilibrium is invariably accompanied by all sorts of disorders before a new unknown equilibrium is established.

The sociologists and biologists use the same vocabulary: mutation, adaptation; but do they understand the same reality by these words? Or rather, are their categories not situated in quite another span of time? The biologist is concerned with a very long series of generations; for the social scientist the concern is for a moment in history, which occupies us no more than an instant in the infinity of time. It is at this moment that we must live and act. Social organization and all human relations rest on a biological substrate. Is it this substrate which is in the process of changing? In what manner, at what rhythm, and with what repercussions? For some, demographic change will require a projection of their thinking over the long term; for others, it means revising their interpretation of events and their perspectives at every moment, taking into account unexpected facts. In an era of very rapid change, it is more important than ever to assess our situation carefully, so that we may make every effort to avoid unforeseen dangers.

## REFERENCES


1. Aries, P.: L'enfant et la vie familiale sous l'Ancien Régime, Paris, 1960.

2. Bernard, J.: L'homme changé par l'homme, Paris, 1976.

3. Dupréel, E.: Population et progrès (1914) dans Deux essais sur le progrès, Bruxelles, 1928.

4. Dupréel, E.: La civilisation, II, Conclusion, dans Sociologie générale, Paris, 1948, pp. 319-392.

5. Fourastié, J.: De la vie traditionnelle à la vie "tertiaire." Recheres sur le calendrier démographique de l'homme moyen Population XIV, 3: 417-432, 1959.

6. Girard, A.: Dimension idéale de la famille et tendances de la fécondité. Comparaisons internationales. Population XXXI, 6: 1119-1146, 1976.

7. Hamburger, J.: La puissance et la fragilité. Essaie sur les métamorphoses de la médecine et de l'homme. Dans L'homme et les hommes, Paris, 1976.

8. Jacob, F.: La logique du vivant. Une histoire de l'hérédité, Paris, 1970.

9. Laslett, P.: Un monde que nous avons perdu (traduit de l'anglais par C. Campos), Paris, 1969.

10. Mead, M.: Le fossé des générations, Paris, 1971. (traduit de l'americain Culture and Commitment: A study of the Generations Gap, 1970).

11. Parsons, T.: La parenté aux États-Unis. Dans *Elements pour une sociologie de l'action* (traduit par F. Bourricaud, Paris, 1955.

12. Riesman, D.: *La foule solitaire: Anatomie de la societe moderne*, Paris, 1964. (traduit de l'americain *The Lonely Crowd*, 1952).

13. Roussel, L.: *Le mariage dans la société française contemporaine*, Paris, 1975.

14. Roussel, L.: *La famille après le mariage des enfants: Étude des relations entre générations*, Paris, 1976.

15. Sutter, J.: Fréquence de l'endogamie et ses facteurs au 19e siècle. *Population XXIII, 2: 303-324, 1968*.

16. Valéry, P.: la crise de l'esprit. Dans *Variété*, Paris, 1924.

17. Valéry, P.: La politique de l'esprit (1932) et Le bilan de l'intelligence (1935). Dans *Variété III*, Paris, 1936.

# New Demographic Patterns and Their Consequences

Nathan Keyfitz

Among the many changes now occurring in industrial societies, demographic changes are not the least important. The discussion here will focus on the disappearance of the classic differentials of mortality and fertility once countries have gone through the demographic transition; the conflict between the values of motherhood and of work; the unprecedented rise in divorce in the 1970s; and the new life cycle brought about by changes in the timing of birth and death.

## DEMOGRAPHIC TRANSITION AND DIFFERENTIALS OF MORTALITY AND FERTILITY

Industrial societies are characterized by low birth rates and low death rates; the route by which they reach this condition is referred to as the demographic transition. Its stages are high births and high deaths, high births and low deaths, low births and low deaths, with a gentle or sudden passing from one stage to the next. The several distinguishable groups in a country do not pass through the stages simultaneously and as the transition proceeds, some groups (for example urban, educated, upper class) are more advanced than others in relation to it. At any given moment the more advanced groups have lower rates of birth and death than the groups that follow them, and these differentials persist as long as the transition does. Children have a different meaning for different groups; they were farm family helpers in the 19th century countryside at ages when they were still school children in the city.

Certainly rural and urban areas are not what they were, either in offering contrasting styles of life and work or in the size of family these induce. In 1880 the U.S. census ratio of children under 5 to women 20 to 44 was 0.845 for urban areas and 1.319 for rural areas; by 1970 it was 0.483 for urban and 0.558 for rural. On the basis of this measure of fertility the ratio was 56 percent higher for rural than for urban areas in 1800 and only 16 percent higher in 1970.

In the later stages of the transition the classic differentials of fertility and mortality began to vanish. Whereas once the richer, better educated, more urban categories of the population had few children and the poor, less educated, and more rural had many, such differences are diminishing in Europe and Northern America. The causes for the change are not far to seek. A major reason was that in the past, although all classes wanted fewer children, only the more affluent knew about contraception, so the differential was great. Now contraception is available to all but a diminishing minority.

Aside from this factor, the disparity between wealth and poverty is not as extreme as it was. With the development of Europe and the United States during the course of this century income differentials have probably shrunk, though the statistical evidence on this point is not easy to unravel. One notes, however (8), that the top 5 percent of individuals had about 30 percent of the nation's disposable income during the 1920s, and less than 20 percent in the late 1940s. Again (8) the median money wage for all male workers in 1939 was $1,112 for whites and $460 for Negroes; by 1970 the ratio was much lower, $8,254 to $5,485. Aside from income distribution, the democratization of health services and of education has lessened class differentials of both mortality and fertility.

## ZERO AVERAGE GROWTH

The declines in the birth rates of the last year or two have changed the demographic prospects of the countries of Europe and America. East and West Germany, as well as Austria, show more deaths than births, so the number of German speakers in the world is declining. Great Britain is in the same condition, and France is approaching it. The United States and Canada still show net natural increase, but this is only because they have a bulge of women at child-bearing ages, reflecting their post-war baby boom.

Aside from any concerns about whether the population is replacing itself, the low birth rate shifts the age-distribution of persons alive at any one time toward the older ages. This shift has many consequences, most of them undesirable: higher costs to the community for old-age security and for medical services, fewer young people in proportion in the labor force, slower promotion for those that are in the labor force, less likelihood of innovation in economic, technical, or artistic fields. However, the age-factor could be drowned out by other forces; the countries of slow growth and many old people are not the least dynamic, as illustrated by a comparison of Sweden with Peru.

Once childbearing comes under the control of individual couples, the aggregate number of children is subject to altogether different laws than when it was a purely biological process. One might have thought that the change would make family size more variable; some people like children, others like travel, whereas others like consumer goods. Yet, in fact, the control works in the opposite way: towards uniformity. Most couples want to have one boy and one girl, or some combination close to this. Few want no children at all and few want as many as four children. When people are able to determine family size there tends to be much less variation than occurs otherwise.

In contrast to the more or less uniform choices made by couples at any one time, the changes over time are substantial. Whether children are regarded as an investment or as consumer goods does not matter for these purposes: more of them are demanded when the economic cycle is high, and fewer, when it is low. However, other factors also enter, and in general we must expect ups and downs in the future. These changes need not be as violent as those of the past 20 years, when United States women went from rates that correspond to 3.8 children in a lifetime to 1.8. Births were 4.3 million at their peak in 1961; in 1976 they only slightly exceed 3 million.

## THE DECLINE OF MOTHERHOOD

In American popular expression no value stands higher than that of motherhood. People still subscribe in principle to the view that motherhood is one of the most wonderful things in the world, and this continues to be the official view of most religions. In fact, however, motherhood takes its place within a hierarchy of objectives, and for many contemporary couples in advanced countries that place seems to be a subordinate one. The first and second child are ranked high, but once these are present the desire for further children is less and less able to compete with the attraction of goods and services perceived as alternatives to children.

People might still wish to have five children as their grandparents did, but two factors weigh against their implementing this wish: 1) children do not fit well into modern production and consumption patterns, 2) the earth's ecology cannot stand them. Even in advanced countries no automatic thermostat exists by which environmental strain leads to fewer children; yet in recent years attention to ecology has coincided with declining births.

The decline is occurring despite the fact that a high proportion of the cost of child-raising is not borne by parents but by the local community or the national state. The collectivity bears all the cost of formal primary and secondary education and much of the cost of higher education; through housing subsidies it bears part of the cost of the accommodation that the larger family requires; through day-care centers it aids the working wife. Parks and playgrounds, insofar as they are more used by children, can be counted as subsidies; the same is true of food stamps insofar as these are differentially used by families with children. Tax discrimination favoring children is found everywhere. Even clothing for children is priced lower per unit of cloth than that for adults; local bus fares are usually differentiated in their favor; many hotels accept children without charge.

From an economic viewpoint the number of children that parents decide to have is a function of these supports. If some of the subsidy were withdrawn there would be fewer children, and the same would be true if natural market forces raised the relative cost of some of the essentials of child-raising, such as food, clothing, and housing. Yet there are more of these commodities per couple in advanced countries with smaller families than when families were larger. This suggests that the shift from children to other kinds of expenditure does not take place in a fixed schedule of preferences, but that preferences themselves are changing.

Concern is voiced for the consequences. Austrians, for example, think of themselves as the carriers of a civilization that has persisted over centuries despite attacks by Turks or Russians in the East, and the French and Germans from the West, and the gain and loss of empires. Now the simple decision of parents not to have children, in the midst of unparalleled prosperity, seems to be a greater threat than the enmity of Germans, Frenchmen, Russians, and Turks.

## FERTILITY REGULATION IN THE INTEREST OF THE LARGER SOCIETY

Insofar as they see the shortage of children as a problem, governments and the public turn naturally to influencing parental decisions through the price system. They tend to think of the present number of children as an equilibrium resulting from the amount of subsidy now in effect, and if so, would not increase in the subsidy, especially in a strategic item like housing, raise the desired family size within broad limits in any desired level? Subsidy should not only reverse the decline of fertility, but enable family size to be "fine-tuned" at will.

It is easy to think of circumstances in which such fine-tuning would prove impotent. Suppose that work became important for its own sake. If the decline of family size were essentially due to women wanting to have careers and obtain status in the world of work, then reducing the costs of having children would be operating on the wrong variable, and a very large burden assumed by the State would have little effect on family decisions. Even if women were driven to work initially by the purely economic motive of the inadequacy of husbands' incomes to provide the material goods that the couples want, the process might be irreversible: once in the labor market the wife could become interested in her work and not want to give it up. Moreover, given the drive to equalize pay between the sexes, the subsidies would have to be of inordinate amounts to offset the wages that a woman foregoes in having a child. For women to forego earning $7,000 per year for five years for the care of 3 million children born per year totals over $100 billion for the United States as the direct labor cost of its children. As women come closer to men's kind of work and levels of pay the cost would rise further. Increase of symmetry between the sexes, with men being expected to take as much time off from work to look after the children as women do would further diminish the tendency to have children.

Any policy based on the assumption that work is painful, and the market salary or wage paid is just enough to offset the pain of staying on the job, so that there is an equilibrium between the pain of working and the pleasure of having children, does not fit modern conditions. A better basis for policy would be to assume that women like to get out of the house, that they are increasingly like men in obtaining their sense of identity from their skill in some kind of paid work. If this is true, then subsidies would not be as effective as convenient and high quality day care centers, availability of challenging part-time jobs whose hours are the same as school hours, and similar ways of enabling women both to work and to raise children. Greater participation of husbands in child-rearing would raise the morale of wives, but government policy cannot easily promote this. There is a general shift away from the large families existing in societies where men decide the number of children in the family, and women have no choice but to raise children.

Here is an example of how an incomplete theory can lead to an ineffective policy. A model in which X is the sole cause of A, verified by a high correlation between A and X in the past, leads to an attempt to control A by means of X. But suppose the effect of X on A is swamped by that of Y. If we act on X and disregard Y, not only are we pushing a lever that has little effect on the engine, but it is even possible that the lever Y tends to move in a way that is negatively correlated with X, so each push on X produces the opposite result to that intended.

Many examples are available to induce modesty in the use of current knowledge for policy-making. France used every possible means to raise her birth rate through the 19th century, with little result.

If there is no direct and simple way of stimulating parents to have children, like raising the income tax exemption, then perhaps coercion rather than economic incentives is what we should be thinking about. Make abortions illegal, prevent wives from working, disqualify men with less than three children for certain jobs. Rumania made abortion illegal and strikingly increased its births, at least for a while. Such measures are time-honored ways of raising fertility.

For good or bad, the path by which we arrived at universal knowledge and use of contraception is not easily traversed in the opposite direction. We came this way on the slogan of voluntary parenthood; giving parents free choice was the reiterated aim of the birth control movement. Taking away the choice offends against deeply felt values, whose inculcation is partly the result of the birth control movement.

The fact that the small family fits into a particular social and environmental context makes any attempt to influence parents quickly difficult, even if one knew which levers to pull. This context may have nothing to do with the equilibrium position, but it makes the response to policy measures very sluggish. If a couple happens to have a large single family home, then they might well have another child with little urging; but if they are living in a small apartment, the cost and trouble of moving makes them disinclined to enlarge their family. An apartment-dwelling couple without children, most of whose friends also have none, are not in the culture of child-raising, with all the informal exchange of information that is so important in the new way of life. Relocating themselves in a new apartment and with a new circle of friends would take time and effort. In such a framework the long-run elasticity of response to a stimulus may be considerable, but the short-run response is low.

## THE SHRINKING HOUSEHOLD

Low rates of childbearing have made the household smaller at the same time as it has been reduced in size by a quite different historical trend: the departure from the household of unrelated persons, as well as related persons other than husband and wife. The old-fashioned small boarding-house has disappeared, and so has the widowed mother-in-law who once occupied the family's spare bedroom. Censuses show the net effect in a drop from 5.79 persons per household in 1790 to 3.58 in 1970 (8). In 1910 5.4 percent of married couples were without their own household, usually living with parents, against 1.4 percent in 1970. Households with subfamilies numbered 2,062,000 in 1940 against only 1,150,000 in 1970.

The same tendencies are reflected in the distribution of households by the number of individuals they contain. Thus one-person households were only 3.7 percent of all households in 1790, a proportion that has climbed steadily to 17.0 percent by 1970. During the same period households with 7 or more persons dropped from 35.8 percent to 5.1 percent. Some 62 percent of households had 2 to 4 persons in 1970, against only 33 percent in 1790. The ideal living arrangement would appear more and more to have become a permanent honeymoon, disturbed at most by one or two children.

In Talcott Parsons' phrase, "The most important single distinctive feature of our family structure is the isolation of the individual conjugal family." Philippe Aries has said of Western Europe: "The modern family . . . cuts itself off from the world and opposes to society isolated groups of parents and children." These long-term tendencies have been accentuated in the recent past.

Of course one must not exaggerate. Hareven (2) warned that the family was not as extended in the old times as we tend to think, and that it is not as conjugal today; however, would-be revisionists of the Parsons and Aries formulation cannot ignore the numbers and trends in the constitution of the household as shown in successive censuses.

This restrictiveness of the household to the small nuclear family contrasts strongly with other cultures. The tragedy of Willems' transgression in Conrad's *Outcast of the Islands* was less his losing a good income than his losing the admiration and servility of the numerous in-laws that filled his house. That the unclean crowd of his wife's relatives had lived by the grace of his will was power, and Willems had loved it.

I have observed in a Javanese village that a man who is at all well-off will often have living with him a widowed aunt, perhaps a cousin or two, and perhaps some unrelated lodgers who are also part servants, as well as his own wife and children. If the lack of privacy ever occurs to him it is more than offset by the feeling of importance at being in a position to support a number of adults as well as his own wife and children. Having parents or in-laws who, by living with him, can thereby witness his success in the world, is a large part of the satisfaction of doing well.

In a traditional society a person would no sooner think of putting his parents in an old-people's home than we would think of putting our own children in an orphanage. Attitudes towards old people have changed from respect to bare tolerance as we have moved from a static agricultural society to a rapidly evolving industrial one. Having no old people or other kin within the household may add to the difficulty of raising children, but with the process of development, people apparently have come to value their privacy. They confine the household to the nuclear family, and have a very small number of children within that family.

## SEX CONTROL OF CHILDREN

A technological factor is likely to make the family of the future smaller yet. Roughly combining information on what people do about having children with what they say is their ideal, we find the objective of many couples to be 1) at least one boy and one girl who would continue the male and female line and 2) a total number of children as small as possible, consistent with this objective. It can easily be shown that if the chance of a given birth being a boy is one half, the mean number of children for at least one boy and one girl is three children. With sex control this would come down to 2 children. If, therefore, a feasible means of sex control becomes available those couples that continue having children until they have at least one boy and one girl will need to have one-third fewer children. No one knows exactly what proportion of couples have so definite a target, but insofar as this is a factor, sex control would substantially reduce the birth rate. The method would have to be simple if couples are to be expected to use it, say no more difficult than present birth control.

Sex control is more likely to find application in developing countries. Particularly in countries that value boys, the use of such a device would ensure a very high male:female sex ratio among births. The immediate effect would be fewer children, by a fraction of one half for those couples who want boys only, ranging downwards to zero for those couples who are indifferent to the sex of their offspring. (We disregard the possibility that ability to have boys would increase births, though this is not excluded logically.) The primary effect of fewer girls would be followed a generation later by a substantial fall in marriages and hence in births, giving sex control a double impact.

A tertiary effect would occur as women became visibly scarce and the sex preference of parents shifted towards girls. Ultimately an equilibrium sex ratio would be attained, possibly after a series of waves. No one knows how quickly the waves would be damped, nor what their periodicity would be. That depends on how quickly the numerical predominance of boys becomes visible to prospective parents and how quickly they respond. It could be 5, 10, or 20 years.

## CHILD-REARING IN AN INDUSTRIAL AGE

What makes it hard to see a reversal of the downtrend in childbearing is the intensive character of a mother's work. Diaper service, prepared baby foods, and similar developments do tend to make the hours of work shorter. However, these are peripheral to a mother's central work, which is nothing less than converting a young organism into a human participant in the particular culture. That has been done successfully billions of times, but it is still no easy task. In most societies it is done by parents who learned from their own parents the techniques for managing and teaching children. A child from a small family has had little chance to learn the art of parenthood by watching his father care for his younger brother, and being assigned this task himself now and again. The correlation between numbers of children in successive generations is known to be positive, and there are many possible interpretations of this, including class differences in fertility. But a part of the reason is that children brought up in small families lack that apprenticeship in parenthood, and having less assurance that they can bring up their children well, are less inclined to risk parenthood.

The risks and difficulties are greater in a society undergoing rapid change. The parent of 1977 fears that the right way to acculturate his child is not the way he himself was taught at his mother's knee. What mother knows best is not what her son or daughter will find most useful. The difference in aims between the generations was manifested in various ways during the 1960s. All the institutionalized certainties of one generation wilted under the disregard or scorn of the next. Would-be parents may well fear that their children will treat them as they treated their own parents.

The discipline to which children are subject in many cultures encourages parenthood. Not only is the parent flattered by having an image of himself in the home, but he is served in many ways, both moral and physical, by his children. A part of the reason for having children must always have been this stimulation of the ego of the parent. In 1977 there are few advanced circles where children make it their business to flatter their parents. Divergence of the peer culture of children from the dominant culture of their parents characterizes our age.

A wife has not learned all she knows at her mother's knee--much of her knowledge comes from school, and schooling is less sex-discriminatory than most other institutions in modern society. Girls and boys are taught much the same things, and the principles of how to handle children, and the satisfactions of having children, are no large part of it. Schools are more oriented to work than to motherhood; notions of efficiency, of technical advance, and of application of scientific knowledge are inculcated. An especially talented girl receives clearly in the course of her prolonged schooling the message that her abilities and training will be wasted as a housewife, and even the mediocre come to feel this. The kinds of skills required for motherhood are given little attention, partly because they do not fit well into coeducational classes; nor under modern conditions is this compensated for by the informal culture of girls' groups in and out of school. These are oriented toward sports, dating, and in some degree, marriage, but very little toward motherhood. Mother may still be a good word, but it is less and less promoted either in the family, the peer group, or the educational system.

## MITIGATING THE INTENSITY OF THE WORK OF HAVING CHILDREN

If it is the intensive work involved in socialization of the child that makes child-rearing so at odds with the pressures of modern society, perhaps some substitute can be found for the labor input. We have automatic devices in industry, machines that tend machines; can we not have machines that aid the housewife in tending her children? The main exhibit here is television. Instead of a one-on-one or even one-on-five relation of mother and child we have a one-on-a-million relation between the speaker on the TV screen and the child. Behind the speaker on the children's program stands a team of producers, artists, actors; instead of a hit-and-miss attempt to find the things that appeal to the child the television producer tests his show or cartoon on a sample of children. Instead of the limited range of stories and other stimuli that the individual mother can provide in between making beds and preparing dinner, television can search the world and range through all cultures to offer an endless variety of stimuli. It is a thoroughly professional surrogate mother for as many hours a day as the child can take.

One would have thought that the outcome would be more knowledgeable, intelligent, alert children compared with earlier generations raised without television. No one knows exactly what the ultimate result will in fact be, for the first generation to be raised by television is still only 25 years old. The outcome could well be more knowledgeable people, but not, as far as present evidence suggests, more highly motivated ones. The one thing that television as a substitute mother cannot provide is a reaction and response to the child. It can provide a surrogate mother only in one direction of the interaction process. The child responds to it, more or less, but it does not respond to the child. Whether this will make more passive adults remains to be seen; that children tend to be stupefied by television is often asserted but not yet known.

In any case television and other household devices take only a fraction of the work out of childbearing.

## CONCENTRATION OF AGES AT CHILDBEARING

One way of reducing the labor cost of raising a family is to have whatever children one is going to have, close together. A mother who has her children spread over her reproductive life, even though there are only three of them, may be occupied as a mother for 25 years or more. If she has them at two-year intervals she can effectively complete her mothering in half that time and then resume her career, or educate herself for some new kind of work. Contraception was referred to as a means of child-spacing in an age when a euphemism was required; now that sure and apparently safe contraception is fully implemented we see that it has rather the effect of child-concentration.

In the United States the variance of ages of childbearing dropped from about 46 to 34 between 1920 and 1970. In France the drop was from about 40 at the beginning of the century to about 32 now (4). These changes are partly the result of fewer children, along with a wish to avoid the dangers of childbearing late in reproductive life. Nearly everywhere that contraception is introduced its first effect is to reduce births in older women. Compression of births occurs, as well as moving them all closer to the time of marriage. Thus, (8) for United States white women the median interval between marriage and first birth fell from 20.3 to 15.5 months between the early 1930s and the late 1960s; the interval from the birth of the first child to the birth of the second child fell from 32.3 to 29.3 months. Similar results appear for nonwhites. Later intervals do not change in any systematic manner, but the conclusion is clear: the modern woman wants to get her child-raising completed early.

Biologists will have more to say on this subject; from a demographic viewpoint youthful childbearing makes for higher turnover, and as far as the rate of natural increase is concerned, partly compensates for the few children born. The concentration of ages at childbearing, along with the removal of the tail of distribution that extended at one time well into the 40s, lessens the damping effect after a disturbance such as a period of low or high births. Insofar as this damping is a biological and social advantage we have to worry about the concentration of births within a few ages. They have a tendency to accentuate the cycles to which modern reproduction tends under economic influences.

Under the conditions of mortality, marriage, sterility, and age of childbearing that exist in developed countries, if all couples that survive and are fertile average three children the population will increase at nearly one percent per year. For bare replacement an average of 2.3 children per fertile couple is required. The above considerations suggest that this is more than couples in rich countries are likely to average in the future. As Ryder has said, "The bearing and rearing of children in a developed society are too expensive for the individual couple, and the rewards of parenthood are too uncertain within the conjugal family design"(7).

The conception, bearing, and rearing of children remain one of the few functions left to the family, which has lost most of its assignments in production, formal education, religious communion, social security, defense, and provision of business credit.

Common activities are what hold groups together (3), and when those of the family are attenuated, divorce can be expected to rise.

## DIVORCE AND INDUSTRIAL SOCIETY

When American divorces increased from 264,000 in 1940 to 385,000 in 1950, it was assumed that this was due to the dissolution of hasty and inappropriate marriages made under unusual wartime conditions, plus the strain of enforced separations caused by the war. But long after the effects of the war were past divorces continued to rise--to 479,000 in 1965, and this was just the beginning. A further acceleration continues to occur in the 1970s, the 1974 figure being 970,000; preliminary statistics for the 12 months ending August, 1976, show 1,083,000 divorces.

The relation of divorce to childbearing is complex. A couple who sees a danger of splitting up will tend to avoid children, who after the breadkup are a handicap both to work and remarriage, whether they remain with the mother or the father; yet there are more single parents than ever today. According to the United States Bureau of the Census (10):

> Estimates indicate that, among persons about 30 years old at present, close to 95 percent have married or will eventually marry, close to one-third will obtain a divorce, and close to four out of every five of those who obtain a divorce eventually remarry. Although women who obtain a divorce eventually have about the same number of children as those with continuing first marraiges, those who obtained a divorce would probably have had more children if they had continuing marriages, in part because they tend to marry about 2 years younger than other women.

Divorce rates and other statistics tell us that marriage is no longer an institution in which family, friends, and neighbors have an interest, influence the choice of mate, and effectively compel couples to stay together. Marriage tends to be unconnected with work, uninfluenced by religion, and separated from security. Instead, in many instances marriage provides companionship and solaces for the bruises inflicted by a competitive industrial order. If it does not achieve this for a couple they have every inducement to try again with other partners. Being the private business of the couple concerned rather than being tied in with other institutions, marriage has become fragile.

Not many years ago the family was a major means of social mobility: the man a girl chose to marry determined both her financial and social standing for life. Similarly for the man, a major question in marriage was whether the girl would help him rise in the world. In those circumstances the choice was so fateful that the ideology of romantic love, a kind of intoxication, was needed in order to enable the young couple to make a decision on the basis of the little knowledge they had either of the prospective partner or of the condition under which such a partner would be a help. In order to make a rational decision, a girl considering marrying a young contractor had not only to know about his capacities, but also about the prospects for the construction industry. That kind of information not being available, all that she could do was to trust her heart. Mistakes could sometimes, but not very often, be rectified by divorce. The young lawyer who married a girl from his home town, and who found that she did not fit the society he had reached might be typical of those divorcing in the days when social mobility predominated.

But social mobility has been attenuated in recent decades, with the boundaries and the very definitions of social status blurred. Causes related to mobility can hardly explain the current rise in the divorce rate. The bourgeois marriage has weakened as property income has diminished in importance, compared with wages. Most pertinent must be marriages coming to depend more and more exclusively on pure affection, rather than on common activities, and less and less tied into other institutions. The high divorce rate is facilitated by legal provisions that make the mere agreement of the couple themselves sufficient for divorce, provisions that would have been thought shocking a generation ago.

Evidently the high and increasing divorce rates are not due to dissatisfaction with marriage; on the contrary, they often result from couples wanting to try again with new partners. The number of persons who report themselves as divorced in the census, which they do only if they have not remarried, is only about 4 percent of the total population 18 years and over, is higher for females, and lower for males.

The symmetry attained in law, by which the rights and freedoms of the two partners are equally protected, does not apply _de facto_. In a society where youth continues to be prized, a young spouse is desirable. In the reshuffle constituted by divorce and remarriage the man almost invariably finds a new wife younger than himself; his ex-wife is more likely either to remain unmarried or to make a less desirable marriage. Equal rights for men and women is a strong value in our society, but sexual inequality in marriage cannot be corrected without violating an infinitely stronger value--the right of free choice of spouses.

We need only a few more years of the present trend to make the couple that stay married through life the exception rather than the rule. Under such drastic conditions, quite possibly tied to some basic features of industrial society, no one can tell what will happen to the family. If both husband and wife follow separate careers, and the husband gets an opportunity to pursue his in some distant place, will this become a usual occasion for the separation of the couple? If the wife is equally intent on her career it may well be. If wives are less intent on careers, and work as typists, nurses, librarians, or elementary school teachers, where they are moveable but not promotable, then the symmetry of the sexes will never be attained.

The choice seems to be between a job that is casual or a marriage that is casual. Only a small minority of the population is yet at this point, but continuation of present trends will increase the proportion. A centralized country like France is in some degree protected against this problem, at least as compared with a thoroughly decentralized country like the United States.

Ryder (7), writing somewhat before the latest sharp rise in divorce rates, predicted that despite its present changes, "because society needs it and because the individual needs it the family will survive." At least for the time being this seems the best assumption to make.

Let us then turn to the life cycle of the individual in the family. Because this has been thoroughly analyzed by Paul Glick (1) the treatment here can be brief.

## THE CONTEMPORARY LIFE CYCLE

Between the beginning of the century and the present time the expectation of life has risen from 47 years to 71. This alone would cause more individuals to reach the time of child-bearing, and also more to survive the growing up of their children. But in addition, marriage is earlier, the couple start having children earlier in marriage, they have their successive children closer together, and they stop with a second or third child, rather than with a fourth or fifth. All these factors together result in the last child getting married at a much younger age of the parents. The decline of the age of marriage helps towards this twice: once in each generation. In addition, customs have changed so that children often leave home and take up residence elsewhere, perhaps at college, before they get married. Whereas at the beginning of the century the couple that survived until their last child left home was in a minority, now most couples see their last child leave.

With this cycle of marriage, child-rearing, and the departure of the children there is a corresponding cycle of geographical movement. The young married couple, usually both working, may start in a small apartment near the center of the city. With the advent of children they need more space, green area, and quality schools. They move to a house in a suburb; the commuting is part of the cost of having children. When in due course the children leave, mowing the lawn and shovelling snow seem an unnecessary burden, and the couple move back to the city. The husband is now at the height of his income: middle-class career arrangements are such that the maximum income comes long after the maximum need to meet family responsibilities; commonly also the wife resumes the work she dropped to rear the children, or else starts some new work. They can well afford a luxury apartment, and move into high-rise, high-rent premises that are now visible in the centers of most large cities. A new demographic element, husband and wife whose children have left home, has its effect on the skylines of many cities.

Thus three distinct phases of the life cycle are separated out: dependency, adulthood, and the child-free ages. Hareven (2) has told of changes in America between the 19th century and today that point the way to the future.

## MIGRATION

An associated aspect of demographic change is migration. In a process now nearing completion in the older industrial countries, people moved into cities as the efficiency of agriculture rose so that they were no longer needed on farms, and as city jobs at better pay called them. The result was extreme crowding in a few places as everyone tried to be at the center. Kirk (6) pointed out that some 70 percent of the U.S. population lives on 2 percent of the land area, and that one-third of counties actually declined in population between 1960 and 1970. Of the few people needed for agriculture with modern equipment some do not even live on the farm, but commute from the city. The same corporations, such as Dow Chemical and Coca-Cola that produce city goods are big agricultural operators.

Within and among cities there is constant movement. In the United States one person in five changes his house each year. People belong to neighborhoods less than they did; they live among strangers and have their community in a profession or in dispersed groups of friends or relatives. These exercise some degree of social control, but not as much as the old fashioned neighborhood, parish, or other local group; and the diminished, informal, local control means that crime has to be kept within limits by the police and the courts. Such a shift of the task of security from the neighborhood, which does it preventively, to the police, who can only act after the fact, seems a part of the cost of the mobility that goes with industralization.

## CONCLUSION

To try to see what is happening demographically in Europe and America is important for several reasons. To assess the future population of these societies requires a judgment on whether the low birth rate now reached is permanent or a mere fluctuation. If the equality of women and their incorporation into the labor force, the independence of children and their unsusceptibility to parental control, and the high rates of divorce are durable conditions, then so is the low birth rate. That this configuration is to be found in some degree in all industrial countries suggests that it is here to stay and that the populations of these countries will at best replace themselves. From the viewpoint of pressure on the environment resulting from both population and wealth, zero increase is desirable. Some countries may fear that their populations are too small to make the contribution to culture of which they would otherwise be capable, but this is a price of the good demographic example they set to the rest of the world.

The underlying conditions of the present low birth rates--geographic mobility, women working at wage jobs, and high divorce rates--will spread to the presently poor countries as these industralize, or even prior to their industralization. Such evidence as exists on birth rates around the world suggests that these modern tendencies are indeed worldwide; if this is so one can be hesitantly optimistic on the possibility that mankind will have dropped to bare replacement early in the 21st century. That would mean, according to our calculations, an ultimate world population of just under 9 billion.

## REFERENCES


1. Glick, P.C., and Parke, R., Jr.: New approaches in studying the life cycle of the family. Demography 2: 187-202, 1965.

2. Hareven, T.K.: The last stage: Historical adulthood and old age. Daedalus 105: 13-27, 1976.

3. Homans, G.C.: The Human Group, New York: Harcourt, Brace and Co., 1950.

4. Keyfitz, N., and Flieger, W.: World Population: An Analysis of Vital Data, Chicago: University of Chicago Press, 1968.

5. Keyfitz, N., and Flieger, W.: Population: Facts and Methods of Demography, San Francisco: W.H. Freeman, 1971.

6. Kirk, D.: The misunderstood challenge of population change. PRB Selection No. 44, Washington, D.C.: Population Reference Bureau, 1973.

7. Ryder, N.B .: The family in developed countries. In The Human Population, San Francisco: W.H.Freeman, 1974, pp. 81-88.

8. U.S. Bureau of the Census: Historical Statistics of the United States, Colonial Times to 1970. 2 vols. Washington, D.C.: U.S. Government Printing Office, 1975.

9. U.S. Bureau of the Census: Statistical Abstract of the United States, 96th annual edition, Washington, D.C.: U.S. Government Printing Office, 1975.

10. U.S. Bureau of the Census: Population of the United States, Trends and Prospects: 1950-1990. Current Population Reports, Series

P-23, No. 49, Glick, P.D. (Ed.). Washington, D.C.: U.S. Government Printing Office, 1974.

11. Westoff, C.F.: The Populations of the developed countries. In The Human Population, San Francisco: W.H. Freeman, 1974, pp. 69-77.

# New Demographic Regime and Its Consequences

Albert M. Jacquard

The demographic history of humanity may be summarized, without too much simplification, by distinguishing three periods separated by two "revolutions," according to traditional terminology, or "catastrophes," according to modern terminology.

During the prehistoric period, the total number of humans on the whole surface of the earth was in the range of one hundred thousand to one million. According to Biraben this number grew from about 700,000 or 800,000 in the year 35,000 B.C. to approximately 7 or 8 million by the year 10,000 B.C. (1). The slow pace of this progression and the small numbers involved in it may be explained by the limitation of the resources available to populations living by hunting and gathering.

The first demographic revolution was brought about by the invention of agriculture. The human population then grew rapidly, increasing to 80 million by the year 5,000 B.C. Then the progression slowed: from 250 or 300 million at the beginning of the Christian era to 800 million at the end of the 18th century.

At this time the second demographic revolution occurred. Success had finally been achieved in the struggle against disease and death, particularly the death of children. Thanks to more effective medicine and, above all, to more sensible hygiene, life expectancy has increased from 38 years for women in France at the end of the 18th century to 76 years today. The proportion of children dying before the age of 15 years decreased simultaneously from 34 percent to 3 percent. The consequence has been a totally new appearance for the human population curve. A slow and apparently linear growth has been followed by an exponential and truly explosive growth: 1.2 billion in 1850, 1.6 billion in 1900, 2.5 billion in 1950, 3.6 billion in 1970, and close to 4 billion today.

At the present rate of increase, three years are sufficient for an addition to humanity of as many humans as were living in the time of Jesus Christ.

This explosion remains an abstract problem for most of us, at least in those countries where the population density is still low or, in any case, tolerable; but it is one of the realities which weighs most heavily on the human condition today.

Let us try to illustrate this point by the following calculations. Since the emergence of Homo sapiens, regardless of the date on which this event occurred, it can be estimated that the number of years lived by the total human population is on the order of 2,000 billion. Persons alive today, our 4 billion contemporaries, have lived or will live for a total of about 200 billion years. In other words, the people living today will accumulate a sum of human experience equal to one tenth the experience lived through by all of the people who have ever existed from the beginning of humanity. If the present trend continues, humanity will live in the course of the 21st century for more than half the number of years it has experienced since its origin.

This exponential growth, which may appear as one event among many, so long as its description in terms of figures masks its reality, can only lead to a catastrophe if collective action is not developed rapidly.

Certainly, opinions may differ considerably as to the maximum number of humans that earth can accomodate and nourish. Serious studies, whose results are naturally quite divergent, have been conducted to evaluate food resources, energy sources and available water. For the purposes of this discussion, we can determine the ultimate limits of human population growth by considering the natural resource which is least expandable: space. The total land area of the earth is 150 million $km^2$. Utilization of mountainous terrain, together with reclamation of the deserts and the frozen wastelands of the Arctic and Antarctic, might conceivably allow us to populate the entire earth at a density equivalent to that of Japan today (300 inhabitants per $km^2$). However, even if this unlikely result were achieved, there would be living space for no more than 45 billion persons. At the present rate of population increase, this limit will be attained before the year 2100.

Let us suppose that an effort is then made to organize settlements on the water, as in certain overpopulated regions today, and that a density of 300 inhabitants per $km^2$ is attained upon the surface of the oceans as well as upon the continents. Such an arrangement would permit an ultimate population of 150 billion. This figure may appear considerable, allowing a certain margin for growth. However, at the present rate of increase, this ultimate limit would be attained shortly after the middle of the 22nd century, less than two hundred years from now.

Disputes concerning the maximal size of the human population are thus quite pointless. Even if estimates differ by tens of billions of individuals, such differences translate into only a few decades of time. A doubling of the maximal population size would only postpone for 35 years, or a little more than a generation, the inevitable date when this maximal size will be attained.

It is evident that the third demographic revolution, the change from an exponential rate of population growth to quasi-stability, cannot be avoided, regardless of our wishes in the matter. The choice can only be between the abandonment of medical gains which help us to overcome disease, and the limitation of births. Few persons will accept the former alternative, so the consequences of the latter must be considered.

In a number of countries occupied by Western civilizations, this revolution has already been realized, a fact which prevents some of their inhabitants from taking the global problem seriously. In some countries, the goal of stability has even been passed and a downward trend has been set in motion. However, these are only local phenomena without any influence on global evolution. There remains only one century for the extension of the new demographic regime to all of humanity, in order for the mean rate of population growth to reach zero.

All available evidence suggests that such a reversal will have innumerable effects on social organization and on individual attitudes. The third demographic revolution will change the mode of transmission of cultures and genetic heritage. We will confine ourselves here to discussion of a very special aspect of the problem, one which is not often brought to our attention: the consequences of international variations in the time of transition to the new regime, and above all the variations in rate of growth once a stable regime has been achieved.

To accept a population growth of zero is to accept a culture in which the right to procreate is subject to extremely severe regulation (as in some Indian states until quite recently), or to social pressure which eliminates deviations from the "normal" behavior (as in current Western societies). Such a culture has characteristics which spring directly from the age structure of the population. The famous "pyramid" of ages is replaced (to draw another image from Egyptian antiquity) with an "obelisk" whose successive age groups have practically equal populations; a sharp decrease in size is observed only in the very oldest classes (more than 80, 90, or 100 years).

Let us illustrate these two patterns by actual examples: The populations of Mexico and of Egypt illustrate the pyramid model; on the other hand, the German Federal Republic and Hungary represent the obelisk model. The age distributions in the two models are shown in the following tables:

Pyramid Model

| Country | 0-14 years | 15-64 years | 65 years |
|---|---|---|---|
| Mexico | 46.2 % | 50.1 % | 3.7 % |
| Egypt | 42.4 % | 54.4 % | 3.2 % |

Obelisk Model

| Country | 0-14 years | 15-64 years | 65 years |
|---|---|---|---|
| German Federal Republic | 23.2 % | 63.6 % | 13.2 % |
| Hungary | 20.2 % | 67.4 % | 12.4 % |

One need only to compare these figures to imagine how different the internal dynamics of these societies must be. Above and beyond any differences in political systems or dominant ideologies, the weight of aged persons, the pressure of the young, and the burden supported by adults are of totally different magnitudes in the two types of societies. The mechanisms and numerical constraints put into effect by the societies are also different; they are necessarily divergent.

The differences in the time at which various societies enter the third demographic revolution, therefore, lead to an international division, the tensions and consequences of which are difficult to forsee. The cultural differences which give to these variations in timing can only be accentuated by the resulting numerical divergence in evolution. This is a process which does not regulate itself, unlike most of those that are encountered in the animate or inanimate world. It is a phenomenon that is truly explosive (or "catastrophic" in the sense referred to by R. Thom), one that our minds are ill prepared to comprehend in its full dimensions. We are accustomed to encountering only autoregulatory phenomena, because phenomena which are not self-regulating cannot persist for long;

thus we can only experience these in a very limited fashion. However, it would be absurd to conclude that such phenomena do not exist or that they should not be confronted.

The time lag in attainment of demographic stability by certain societies has obvious consequences for the population sizes of various human groups. The current rate of increase in the number of humans, about 2 percent per annum, represents a mean between populations such as those of Europe, which have already achieved stability, and populations with a growth rate of more than 4 percent.

Thus the populations of France and of Mexico were equal in 1970: 51 million for the former, 50 million for the latter. However, the annual rate of growth in Mexico was 3.4 percent with a tendency toward acceleration (the rate was only 1.1 percent in 1930 and 2.7 percent in 1950). In France, on the other hand, the rate of growth was less than 0.5 percent in 1970 and was decreasing toward zero. Assuming that these trends are maintained until the end of the century, it can be calculated that the population of Mexico will exceed 200 million by the year 2000, whereas the population of France will be only 60 million. A lag time of one generation is thus sufficient to change the population ratio from 1:1 to 4:1.

These reversals definitely have implications for the power of relationships between groups. Power does not depend on population alone, but population pressure could give rise to defensive reactions on the part of those who feel that this component of their power is gradually slipping away.

One of the consequences, occasionally mentioned, of this quantitative evolutionary divergence is the transformation of the collective genetic patrimony of humanity: genes are quasi-immutable, eternal; they multiply themselves in identical fashion, as the number of descendants increases. To return to the preceding example, the "Mexican genes" will represent in the year 2000 a portion which will be four times as great as that of

"French genes" in the biological patrimony of humanity (2).

At first glance it may be observed that this evolution is difficult to describe as favorable or unfavorable, given the impossibility of assigning any kind of "quality" to most genes.

However, we must go a little further and state that this genetic modification of humanity, no matter how high the figures by which it is measured, is practically insignificant. In order for these changes to be of real consequence, it would be necessary that the Mexican genes and the French genes be different, so that in general the various human groups would represent genetically distinct races. Certainly most genes have variable frequencies from group to group; some are even absent in certain groups and highly prevalent in others. However, on the whole, those conspicuous variations which apply to readily apparent characteristics, such as the color of the skin or the texture of the hair, represent only a small part of the genome as a whole. Recent analyses by Nei (5) and Lewontin (4) have attempted to estimate the portion of the total genetic variability of humanity due to differences among individuals of the same group as opposed to that due to differences among the means of groups. This analysis of variance, which was fortunately limited to a few dozen loci for lack of more extensive information, reveals that variations between groups represent no more than 7 or 8 percent of the total variability. Thus a modification in the weight of the various ethnic groups scarcely modifies the global genetic structure of our species.

The geneticist could, therefore, feel little concern about this inevitable reversal. When he examines the more detailed effects, at the level of the age of procreation or the variance in the size in families, he may definitely produce evidence of certain genetic consequences for the future order, but these are of little importance; they have divergent effects which more or less counterbalance each other (3).

The real consequences do not concern the genetic patrimony of humanity, but rather its culture; the effect is not upon those characteristics which are inborn, but upon those which are acquired. The following table shows the distribution of the human population by continents at the beginning of the Christian era and in 1970:

| Time | Continent: Asia | Europe (including the Asiatic part of the U.S.S.R.) | Africa | North and South America |
|---|---|---|---|---|
| 1 A.D. | 65% | 20% | 11% | 4% |
| 1970 A.D. | 57% | 19% | 10% | 14% |

Thus the changes are of little importance in the final analysis. Only America, which was practically empty for 2000 years, has increased its share of the population, to the detriment of the continent of Asia. The exact outcome is due to a great variety of historical circumstances which, strangely enough, have given rise to a distribution which closely resembles the original one in terms of percentages if not in the absolute numbers. However, this represents an accident of very brief duration. We can illustrate this point by the following exercise: If we estimate the future distribution of population by extrapolating the trends observed between 1950 and 1970 we obtain the following annual rates of growth:

| Asia | Europe | Africa | The Americas |
|---|---|---|---|
| + 1.9% | + 1.1% | + 2.4% | + 2.2% |

This pattern of growth rates would, for each generation that it remains in effect, result in the Asian share of the world's population remaining constant, while the African share would increase by 2 percent and the American by 1 percent; the

European portion would decrease by 3 percent. Of course this is not a forecast but only a translation, in terms of distribution, of the progressive effects of variations in rates of population growth.

At the national rather than the continental level, the disparities are much more pronounced. We have seen that Mexico now has an annual rate of growth of 3.4 percent. At this rate, its portion of the population of the world will be multiplied by a factor of 1.5 for each generation, even though the world population as a whole is expanding.

To be sure, the true weight of a human group is not determined entirely by its population size. However, future modifications in this factor will necessarily lead to a reversal of cultural equilibria, or perhaps to a greater and greater emphasis on modes of political pressure that are independent of population size and therefore, by definition, nondemocratic.

## REFERENCES

1. Biraben, J.N.: Essai sur l'évolution du nombre nombre des hommes. Communication au Congrès de l'U.I.E.S.P., Mexico, August, 1977.

2. Jacquard, A.: _Concepts en génétique des populations_. Masson, Paris and New York, 1977.

3. Jacquard, A., and Ward, R.: The genetic consequences of changing reproductive behaviour. _J Hum Evol 5: 1976_.

4. Lewontin, R.C.: _The Genetic Basis of Evolutionary Changes_. Columbia University Press, New York, 1975.

5. Nei, M.: _Molecular Population Genetics and Evolution_. North-Holland Publishing Company, Amsterdam, 1975.

# *Mutation and Enzyme Variation in Human Populations*

Harry Harris

What is currently known about the mutations which result in inherited variants of human enzymes and other proteins has mainly come from two quite different approaches. One approach has its origin in the classical work of Archibald Garrod on the "Inborn Errors of Metabolism." It takes as its starting point particular diseases which appear to be attributable to single gene mutations and aims to identify in each the specific enzyme or protein defect which gives rise to the observed metabolic and clinical abnormalities. The other approach involves a quite different strategy. It takes as its starting point randomly selected individuals in a population and aims to identify differences between them in particular enzymes or proteins and determine their genetical basis.

The extent and diversity of enzyme and protein variation in human populations which has been revealed by such studies has been quite remarkable and is in a sense very unexpected because it would hardly have been anticipated by most people working in this area just a few years ago. Looking at the data as a whole a number of interesting and novel generalizations have begun to emerge. It is on these new generalizations that I would like to focus.

## How many alleles occur?

The inherited variants of different enzymes and proteins that we find among living members of our species today can be attributed to specific gene mutations which occurred in single individuals among our ancestors in earlier generations--some living perhaps relatively recently, others in the more remote past.

Mutations appear to be essentially random events which occur at a low but steady rate. Furthermore, we can expect that a very large number of structurally different mutant alleles may be generated by separate mutational events within the confines of a single gene. Thus, from a typical gene containing a DNA sequence of say 1,200 bases and coding for a protein of 400 amino acids, as many as 3,600 different alleles may be generated simply from mutational events which result in no more than a single base change at some point in the sequence. This is so because each base may be changed to one of three others. In addition, it is known that other types of mutation involving small deletions, duplications, frame shifts, etc. within the confines of a single gene can also occur, and each will generate its own structurally distinct mutant allele.

So the first questions which arise are, how many of these different possible mutant alleles actually occur among living members of the species, and what are their functional effects?

We can perhaps approach these questions most simply by thinking about the variants of hemoglobin, since this protein has been investigated more extensively from this point of view than any other.

Consider the gene which codes for the beta polypeptide chain of hemoglobin (Hb). There are 146 amino acids which are coded by a DNA sequence of 438 bases; so point mutations each resulting in only a single base change could generate 1,318 different alleles at this locus. From the known amino acid sequence and the genetic code one can classify their effects approximately as follows:

1. About 23.4% (307 alleles) will be so-called synonymous mutants which involve the alteration of a codon specifying one particular amino acid to another codon specifying the same

amino acid. Such mutants cause no alteration in the protein structure.

2. About 3.7% (49 alleles) will be chain terminating mutants, which involve the alteration of a codon specifying a particular amino acid in the polypeptide chain to a so-called "nonsense" codon which specificies chain termination. Such mutants can be expected to result in the synthesis of an abbreviated polypeptide chain which lacks the series of amino acids normally coded by the base sequence in the gene beyond the point where the mutational change occurred. In most cases such an abbreviated polypeptide is unlikely to give rise to a stable functional form of the protein.

3. About 72.9% (958 alleles) will result in the alteration of a codon specifying one amino acid to a codon specifying another and thus lead to the synthesis of an altered polypeptide chain which differs from the original by a single amino acid substituion.

The screening technique which has been most widely and systematically applied to searching for Hb variants is the method of electrophoresis. This will in general be expected to detect amino acid substitutions which involve a change in charge, i.e., substitutions of a neutral for a basic or acidic amino acid or vice versa, and substitutions of a basic for an acidic amino acid and vice versa. One can estimate that 319 of these possible 958 variants would involve substitutions of this sort (i.e., almost exactly one-third). Thus on purely theoretical grounds we can expect that about 319 different electrophoretic variants might be generated by single base change mutations in the $\beta$Hb gene.

Actually among all the Hb variants identified by 1976 (2) some 104 appear to involve this type of amino acid substitution in the $\beta$ chain. Thus we can say that of all the possible mutants of this sort that might exist, approximately 33% are actually known to occur.

Although electrophoretic screening of Hb and the necessary amino acid sequencing of variants has now been carried out very extensively both in patients with hemolytic disease of various sorts, and in random population surveys, it is clear that only a minute fraction of the world's population has been examined in this way. So the fact that as many as 33% of the possible examples of this type of mutant have already been found is quite remarkable. It suggests, indeed, that a very high proportion of all possible mutants actually exists somewhere in the species, though, of course, the majority are likely to be very rare.

The effects of these different mutations on the functional properties of the protein may be expected to vary widely. In the case of mutations resulting in a single amino acid substitution, it will depend on the particular site in the three dimensional structure at which the substitution occurs and on the chemical properties and size of the side chain of the substituted amino acid. The Hb $\beta$ gene mutations again illustrate the point. Different amino acid substitutions give rise to sickle cell and Hb C disease, others to the condition known as hereditary methemoglobinemia, still others to a variety of forms of chronic nonspherocytic hemolytic anemia due to so-called unstable Hbs; and there are others which do not appear to result in any functional abnormality at all. Thus a wide range of biochemical, metabolic, and clinical consequences may arise because of different mutations within the confines of a single gene.

Another important point is that from family studies it appears that only a very small fraction of these variants can be attributed to fresh mutations which occurred in the immediately preceding generation. Most have probably been around for several generations, and many for longer.

There is no reason to think that the Hb $\beta$ gene locus is in any way peculiar in these respects. For example, more than 80 different allelic variants of the enzyme glucose-6-phosphate dehydrogenase (G-6-PD) have already been identified (10,1) and here again, the metabolic and clinical consequences of these different mutants vary widely. Some inevitably result in chronic hemolytic anemia, others appear to have no clinical effects unless the person affected takes a particular drug such as primaquine or one of the sulfonamides or eats fava beans, and others appear to have no adverse effect at all.

A multiplicity of allelic variants strictly analagous to those found for hemoglobin and G-6-PD have also now been demonstrated in studies on quite a number of other enzymes and proteins. Thus there seems doubt that the phenomenon is, indeed, a general one.

I would like to suggest as our first generalization that, at virtually all loci coding for enzyme and protein structure, a very large number of the different possible mutant alleles which may be generated by separate single gene mutations, actually exists among living members of the species. Furthermore, for any particular enzyme or protein the allelic variants which occcur may vary widely in their properties and their functional consequences.

## Polymorphism

This leads us to the question of the incidence of these alleles in human populations. Clearly the degree of allelic diversity among members of a particular population will depend not only on the number of different alleles which may be present at any particular locus coding for an enzyme or protein and on the functional effects of their products, but also on the actual frequencies with which these alleles occur in any given population.

In fact, it turns out that the great majority of these mutant alleles are very rare, but there are some which occur sufficiently frequently as to give rise to the phenomenon usually referred to as enzyme or protein polymorphism. The term polymorphism in genetics is generally reserved for situations where a population of individuals can be sharply classified into two or more distinct phenotypes in terms of some particular enzyme or protein, and where each of the different phenotypes occurs quite commonly in the particular population being studied.

It is necessary, of course, for analytical purposes to have a firm numerical definition of what one is going to call a polymorphism. In what follows, I will use this term for any situation where the commonest detectable allele at the particular locus has a frequency no greater than 0.99. This means that at least 2% of the population will be heterozygotes. In many cases, of course, the incidence of heterozygotes is greater than this. Loci where the commonest allele has a frequency greater than 0.99 will be referred to as nonpolymorphic. This does not mean that they are entirely monomorphic in the strict sense, because probably in most such cases multiple rare alleles do occur in the population and produce detectable enzyme or protein variants; but the population of individuals heterozygous for such alleles is, at most, less than 2% and usually very much less than this. These definitions are somewhat arbitrary but turn out to be convenient and meaningful in practice.

The distinctions are well illustrated by the example of the $\beta$ chain hemoglobin variants, which I discussed earlier. Here we know that while the majority are exceedingly rare and many have only been found in single individuals or among the members of circumscribed family groups, there are a few which are sufficiently common in certain populations as to give rise to typical polymorphisms (e.g., Hb S in Africa and Hb E in Southeast Asia). So in certain populations in Africa and Southeast Asia where the incidence of heterozygotes is often 10-20% or greater, the Hb $\beta$ gene locus is referred to as polymorphic; whereas in other populations where common alleles at this locus do not occur, and the proportion of individuals heterozygous for one or another of the many different rare alleles is usually less than 1 per 1,000, we refer to the hemoglobin locus as nonpolymorphic.

The phenomenon of genetic polymorphism has, of course, been recognized for a long time. Perhaps the best known examples in man, apart from the sickle cell and other hemoglobin polymorphisms, are the series of blood group polymorphic systems which were identified because of the occurrence of common individual differences in certain red cell antigens. Because of the apparently rather special features of these polymorphisms and the manner by which they had come to be discovered, it is not known whether they represented special and rather unusual cases, not typical of gene loci in general; or whether they simply happened to be rather striking examples of a widespread phenomenon.

However, some ten years ago my colleagues and I in London working on man (3), and Lewontin and Hubby in Chicago working on *Drosophila* (8) set out, quite independently, to try to answer this question. The idea, in both cases, was to see whether if one examined by electrophoresis a more or less unselected series of enzymes or proteins in a naturally occurring population of individuals from the given species, one would find any examples of polymorphism, and if so, at what proportion of gene loci the phenomenon occurred. Electrophoresis was chosen because it was known that it could detect mutants with single amino acid substitutions provided these resulted in a change of charge, and in theory about one-third of all mutations causing amino acid substitutions should be of this sort. Also, it is a method quite generally applicable to a wide range of enzymes and proteins and readily adapted for population surveys.

To cut a long story short, it was found using this approach that enzyme and protein polymorphism both in man and *Drosophila*, is a very much more common phenomenon than had previously been expected by most people. Similar studies have now been carried out on many different species and it is clear that such polymorphisms are ubiquitous in naturally occurring populations.

My colleague, Dr. D. A. Hopkinson, and I have recently reviewed the available data on enzyme polymorphisms in man (5). We found that satisfactory information is now available on at least 104 different loci coding for enzyme structure. Of these, 33 loci appear to be polymorphic for electrophoretic variants in at least one major human ethnic group. However, there are differences from one population group to another. For example, among Whites of European origin, only 24 of these loci appeared to be polymorphic and of these, at least two were not polymorphic in Blacks. Similarly there were several loci evidently polymorphic in Blacks but apparently nonpolymorphic in Whites.

The data on the Whites happens to be much more extensive and complete than on any of the other populations and this allows one to obtain an estimate of the so-called average heterozygosity per locus, which is a useful index of this type of variation. Our estimate of this parameter was 0.063. This implies that any single individual is likely to be heterozygous at about 6% of his gene loci coding for enzyme structure, for alleles which result in electrophoretically detectable variants. This, for various reasons which I will discuss later, is almost certainly a serious underestimate of the total heterozygosity. Nevertheless, even as it stands it implies a remarkable degree of individual diversity, and it is not difficult to show from such data that any two individuals with the exception of monozygotic twins are most unlikely to be exactly alike in their enzyme and protein constitutions.

One might expect that this uniqueness of the individual in his basic enzyme makeup would be reflected in his metabolic and physiologic characteristics and also in his degree of susceptibility to the development of particular sorts of disease or other abnormalities. So far, however, it has not proved possible to analyze this question in any systematic way, although it has been shown that in more than half the known cases of enzyme polymorphism, there are significant differences of levels of activity between the enzyme phenotypes which make up the polymorphism. So much for the polymorphisms determined by the relatively common alleles at these various gene loci.

## Rare Alleles

Now let us turn to the question of the so-called rare alleles which determine rare enzyme variants. Many examples of such rare variants have been reported in the literature, but little consideration has been given in the past to their incidence and distribution.

In searching for polymorphism in any given population, the number of different unrelated individuals examined need usually amount to no more than a hundred or two, because the alleles involved are, by definition, relatively common; but to study rare alleles, very much larger samples are required.

The point is illustrated by our own findings in the case of one of the loci which determines the enzyme phosphoglucomutase. Some years ago we showed that three common phenotypes could be identified readily and these were due to two common alleles with frequencies in European populations of about 0.76 and 0.24, the homozygous phenotypes occurring in about 56% and 6% of the population respectively and the heterozygous phenotype in about 36% of the population. Subsequently in the course of examining this enzyme in a large number of different individuals a number of rare phenotypes were identified (6) which could be attributed to the occurrence of rare alleles in heterozygous combination with one or the other of the two common alleles, but only about one in a thousand people was found to be heterozygous for one or another of them.

Similar rare variants had also been detected in the course of investigation of other enzymes, and eventually we decided to put together the data which had accumulated over a period of about ten years in order to see whether we could obtain a general picture of the incidence of such rare alleles over a wide range of loci (4).

We found that we had reasonably reliable data on 43 different enzyme loci. The number of unrelated individuals examined varied for the different enzymes from just a few hundred to up to more than 10,000, but the average per locus was about 2,500, and since the loci are probably all autosomal, this implied that on average we had screened the products of about 5,000 genes at each locus for electrophoretic variants. For the purpose of the analysis we defined a rare allele as one which had a gene frequency of less than 1 in 200 in the general population. In all, 56 different ones had been identified, and more than 80% of these appeared to have frequencies of less than 1 in 1,000, so the majority of these alleles are indeed extremely uncommon.

The tabulations of the detailed findings over the 43 loci are rather extensive, so in order to illustrate the main features of the data, I have simply summarized the results in just 12 of them (Table 1). The first line, for example, gives the findings for the phosphoglucomutase locus ($\underline{PGM}_1$) which have already been mentioned. The number of unrelated individuals tested was 10,333, i.e., 20,666 genes. Five different rare alleles were identified and these occurred in 12 individuals who were heterozygous for one or another of them, so the combined incidence of rare heteroygotes was 1.16 per 1,000. At this locus there are also two common alleles, and heterozygosity in the population due to these is 0.366 or 366 per 1,000.

In the surveys one or more rare alleles were detected at just over half the 43 loci studied. Obviously, the chance of detecting such rare events will depend on the size of the sample. Indeed, the number of individuals examined was on the average much lower among the loci where no rare alleles were detected (1,300), than among the others (4,023). Also at several of the loci, although we failed to find rare variants in these surveys, such variants have in fact been found in other studies; so in general, it seems not unlikely that such rare alleles actually occur at virtually all loci.

It is obvious that estimates of the frequencies of individual alleles in such surveys must inevitably be very imprecise because most of them are extremely rare. It is more practical, therefore, in comparing different loci to consider

Table I

Incidence of rare alleles determing electrophoretic enzyme variants in unrelated Europeans. Further details and also detailed data on 31 other loci are given in Reference 8.

| Enzyme | Locus | No. of unrelated individuals tested | No. of different rare alleles | No. of rare hetero-zygotes | Hetero-zygotes for rare alleles (per 1,000) | Hetero-zygotes for common alleles (per 1,000) |
|---|---|---|---|---|---|---|
| Phosphoglucomutase | $PGM_1$ | 10,333 | 5 | 12 | 1.16 | 366 |
| Phosphoglucomutase | $PGM_2$ | 10,333 | 3 | 7 | 0.68 | --- |
| Glutamate-oxalate transaminase 'soluble' | $GOT_s$ | 1,195 | 2 | 2 | 1.67 | --- |
| Glutamate-oxalate transaminase 'mitochondrial' | $GOT_m$ | 1,195 | 0 | 0 | --- | 33 |
| Adenylate kinase | AK | 6,760 | 1 | 1 | 0.15 | 77 |
| Nucleoside phosphorylase | NP | 1,542 | 2 | 2 | 1.30 | --- |
| Pyrophosphatase | PP | 2,190 | 0 | 0 | --- | --- |
| Alkaline phosphatase-placental | PL | 3,244 | 10 | 76 | 23.43 | 502 |
| Peptidase-B | PEPB | 7,041 | 3 | 15 | 2.13 | --- |
| Triosephosphate isomerase | TPI | 1,750 | 2 | 2 | 1.17 | --- |
| Superoxide dismutase | $SOD_A$ | 11,237 | 1 | 7 | 0.62 | --- |
| Inosine triphosphatase | $ITP^A$ | 641 | 0 | 0 | --- | --- |

the combined incidence of heterozygosity due to the rare alleles at each of the loci. These heterozygosities per locus turn out to vary considerably, extending over a 150-fold range even if we consider only those loci at which rare alleles were, in fact, discovered. To some extent this variation no doubt occurs for technical reasons, because we know that there is much variation in the discriminative power of the electrophoretic methods available for the different enzymes. Nevertheless, we are inclined to think that the variation does to a significant degree reflect real differences between loci.

A case in point is the locus PL which determines a specific alkaline phosphatase peculiar to the placenta and which shows a quite remarkable degree of allelic variation. In the course of examining placentae from 3,244 individuals, evidence for as many as 14 different alleles was obtained. Ten of these can be counted as rare alleles according to our definition and they account for as many as 23 per 1,000 heterozygotes. The common alleles give rise to some 500 heterozygotes per 1,000 individuals. The number of different rare alleles found in this particular case and the degree of heterozygosity to which they give rise, makes us suspect that at this locus some unusual process may be responsible either for generating the allelic diversity or maintaining it at this high level.

Despite this variation in heterozygosity from locus to locus, it is nevertheless of some interest to estimate the average heterozygosity per locus due to rare alleles, over the full range of loci including those where no rare variants at all were discovered. This average is 1.76 per 1,000 for all the loci, but it is reduced to 1.14 per 1,000 if we exclude PL. Thus the data as a whole suggest that on average for any single locus we may expect that between 1 and 2 individuals per 1,000 will be heterozygous for a rare allele determining an electrophoretic variant. If we suppose, for example, that there are 30,000 loci coding for enzyme structure, each of us is likely to be carrying at least 30 such alleles, and it is extremely improbable that any two of us will, with the exception of monozygotic twins, have exactly the same combination; so these rare alleles make a significant contribution to the overall variation between individuals.

These studies were carried out on a European population. Recently Neel and his colleagues (9) have done similar studies in Japan and have obtained essentially similar results.

The alleles which determine these rare electrophoretic enzyme variants must have originated by mutations occurring in single individuals in earlier generations. The question, therefore, arises as to how many of the rare variants observed in these surveys are the consequence of fresh mutations, that is to say, mutations which occurred in the germ line of one of the parents of individuals found to have the variant.

In the course of the work we had systematically carried out studies on the families of individuals showing rare variants whenever this was practical, and an extensive body of pedigree data had been assembled. In fact, no clear example of a fresh mutation was found, but of course, not all the family data is informative, because sometimes critical individuals could not be tested. However, in 77 unrelated individuals who were heterozygous for one or another of these rare variants, we had been able to test both the parents. In each case, either the father or the mother of the individual with the variant showed the same variant, so none of these cases was due to a fresh mutation. One can, therefore, conclude that the great majority of the rare variants observed are not the products of fresh mutations, and the fraction attributable to fresh mutations must be very small--probably less than 1 to 2%.

The findings are, of course, in complete harmony with those mentioned earlier for the hemoglobin and the G-6-PD loci. To generalize, we can perhaps say that:

1. In any given population a multiplicity of different rare alleles probably occur at most loci, both at those that also have two or more

common alleles and are, therefore, polymorphic and at those which do not and are classified as nonpolymorphic.

2. That only a very small proportion of these rare alleles are due to fresh mutations which occurred in the immediately preceding generation.

3. That although individually each is very uncommon, together they make a significant contribution to the biochemical diversity among individual members of the population.

## Alleles Not Detected by Electrophoresis

It is, however, important to recognize that the findings in these surveys both about the number of different alleles and also the amount of heterozygosity at the various loci inevitably underestimate the true values. There are several reasons why this is so.

1. We know from our experience with the electrophoretic systems we have been able to develop that their discriminative power varies considerably from enzyme to enzyme. It is, therefore, quite probable that at some loci we have failed to detect electrophoretic variants which actually occur because of inadequacy of the techniques.

2. We have for operational reasons assumed that electrophoretically indistinguishable variants are attributable to the same allele, but this is not necessarily true, since two quite different amino acid substitutions in an enzyme protein can produce the same change in electrophoretic mobility. So the number of different alleles recorded in such surveys must be a minimal estimate.

3. Because of the nature of the technique, we have only detected variants showing electrophoretic differences, but it is clear that many other mutant alleles could occur which, while modifying the structure of the enzyme, do not alter its charge and hence, its electrophoretic mobility. Indeed, it seems from theoretical calculations that no more than about one-third of

all the possible variants involving single amino acid substitutions would be detectable by electrophoresis.

4. Underestimation must also have occurred for another and perhaps rather less obvious reason. The electrophoretic techniques depend on the use of specific staining reactions designed to detect the activity of the particular enzyme under investigation. Consequently, a variant will only be detected in heterozygous individuals if it has retained the specific catalytic activity characteristic of the enzyme. It is known, however, that certain mutant alleles result in a complete loss or a very marked reduction in the enzymic activity. This may be because the presumed amino acid substitution has affected the catalytic site, or has rendered the enzyme molecule excessively unstable so that it decays very rapidly, or perhaps because it has a reduced rate of synthesis. In a heterozygote for such an allele, one would, therefore, only see the electrophoretic band corresponding to the normal allele with which it is in heterozygous combination, and although this may be somewhat weaker than usual, the methods are, in general, so insensitive to quantitative variation that in virtually all cases the anomaly would be missed.

Evidence for the occurrence of heterozygotes for mutants resulting in such activity loss--so-called "silent" or "null" alleles--has indeed been obtained at a number of the loci which have been studied by these electrophoretic techniques. This evidence has mainly come from the anomalous segregation patterns which are occasionally produced in families in which such an allele is segregating. It has, however, not yet proved possible to obtain satisfactory estimates of their incidence from the population surveys, even though these have been combined with quite extensive family studies.

There is another quite different source of information which tells us something about the incidence of such "null" or enzyme deficiency-causing alleles. This comes from work on the rare, inherited metabolic disorders classically known as the inborn errors of metabolism--the second approach to population diversity which I mentioned at the beginning.

These are conditions mainly inherited as autosomal recessives, in which there is some characteristic metabolic disturbance due to a gross deficiency of a specific enzyme. Many inborn errors of metabolism have now been identified and the important point for our present purpose is that the mutant alleles which determine them are effectively the same as the so-called "null" or "silent" alleles discovered as a result of the anomalous segregation in particular families, of enzyme phenotypes detected electrophoretically.

One well-known condition of this sort is phenylketonuria. Here there is a gross deficiency of the enzyme phenylalanine hydroxylase and this results in a block in phenylalanine metabolism, which, if left untreated, causes severe mental deficiency. It is, however, possible to treat the condition very effectively by feeding a phenylalanine-restricted diet; but to do this it is necessary to diagnose the condition very shortly after birth because if the introduction of the diet is delayed, irreversible brain damage will occur. To cope with the situation it was necessary to devise simple but specific methods for screening all newborn infants in order to pick out those with phenylketonuria. As a result of such surveys, good estimates of the incidence of this disorder have now become available. Furthermore, this general approach has been extended to the screening of the newborn for other inborn errors of metabolism which, like phenylketonuria, can be identified by the presence of very abnormal concentrations of some particular low molecular weight metabolite in blood or urine. Thus, estimates of the population incidence of a number of other rare conditions due to specific enzyme deficiencies are also being obtained.

The most extensive data presently available is probably that collected by Levy and his colleagues in Massachusetts (7) where the great majority of all newborn infants are now routinely screened for a series of different metabolic disorders. Table 2 summarizes some of the results. The number of newborns screened for the various conditions varies from about 330,000 up to nearly a million, and the estimated incidence of the different conditions at birth varies from about 1 in 15,000 up to about 1 in 300,000. The conditions listed are probably all autosomal recessives, so approximate estimates of the gene frequencies can be obtained by taking the square roots of the case frequencies, and from these one can obtain estimates of the incidence of heterozygotes at each of the loci. It seems that about 11% of the population are likely to be heterozygous for an allele which determines one or another of these conditions.

For technical reasons it has not yet proved possible to examine the enzymes involved in this series of conditions by the electrophoretic survey techniques. Nevertheless, it is of some interest to compare the level of heterozygosity obtained by the two different approaches. In fact, the heterozygosities per locus for these rare deficiency alleles are on average several times greater than those obtained for the rare alleles determining functionally active enzyme variants detected in our electrophoretic surveys. This is at first sight a very paradoxical result because in general one would expect that alleles causing gross enzyme deficiencies would be those which were under the most pressure from natural selection, and, therefore, the least common. The discrepancy, however, is probably mainly due to the sampling procedure involved in such surveys for metabolic disease. One does not know how many other metabolic disorders due to deficiency alleles at other loci there might be which could have been picked up in the screening tests but which were not, simply because the allele frequencies in the population were too low. The findings do suggest, however, that there are likely to be many other so-called deficiency or "null" alleles at different loci with frequencies similar to those observed for the rare alleles detected in the electrophoretic surveys.

What we need is some general procedure analogous

Table 2

Incidence of certain metabolic disorders among newborn infants in Massachusetts (10).

| Disorder | Total Screened | Number detected | Incidence | Estimated frequency of heterozygotes (per 1,000) |
|---|---|---|---|---|
| Phenylketonuria | 1,012,017 | 66 | 1: 15,000 | 16 |
| Cystinuria | 350,176 | 23 | 1: 15,000 | 16 |
| Hartnup disease | 350,176 | 22 | 1: 16,000 | 16 |
| Histidinaemia | 350,176 | 20 | 1: 17,500 | 15 |
| Argininosuccinic acidaemia | 350,176 | 5 | 1: 70,000 | 8 |
| Galactosaemia | 588,827 | 5 | 1:118,000 | 6 |
| Cystathioninaemia | 350,176 | 3 | 1:117,000 | 6 |
| Maple-syrup-urine disease | 872,660 | 5 | 1:175,000 | 5 |
| Homocystinuria | 480,271 | 3 | 1:160,000 | 5 |
| Hyperglycaemia (non-ketotic) | 350,176 | 2 | 1:175,000 | 5 |
| Propionic acidaemia acidaemia (ketotic hyperglycinaemia) | 350,176 | 1 | <1:350,000 | 3 |
| Hyperlysinaemia | 350,176 | 1 | <1:350,000 | 3 |
| Rickets vit. D dependent (with hyperaminoaciduria) | 350,176 | 1 | <1:350,000 | 3 |
| Fanconi syndrome | 350,176 | 1 | <1:350,000 | 3 |

to electrophoresis which would enable us to detect such alleles in heterozygotes in population surveys.

## Concluding Remarks

A new and interesting picture of inherited variation in human biochemistry is beginning to emerge from these different approaches.

It seems that at each gene locus coding for a particular enzyme a large array of different mutant alleles occur among the individual members of human populations. The majority are individually very rare but some occur with a significantly high frequency so as to give rise to the phenomenon of polymorphism at a significant proportion of loci.

These alleles exhibit a wide spectrum of functional effects. At one extreme there are those which give rise to a complete or nearly complete loss of the functional activity of the particular enzyme so that in homozygous individuals obvious clinical abnormality results. At the other end of the spectrum only minor functional effects occur or perhaps in most environmental circumstances, none at all.

Thus studies on the biochemical basis of rare inherited diseases complement the studies of enzyme differences between normal healthy members of the population, and together such studies provide us with a new insight into the biological basis of human diversity.

These two aspects of the subject, which for historical and practical reasons have been approached in quite different ways, can now be seen to represent simply different facets of the same fundamental phenomena. They are the two sides of the same coin.

Many problems still remain. We have as yet, for example, little idea of the detailed biochemical and metabolic consequences of most of the enzyme and protein variants which have been discovered; nor do we have any clear idea, in most cases, of exactly how most of the allelic variants which occur have actually come to assume the frequencies in populations that we actually observe. How far, for example, have these frequencies been determined by natural selection and how far by random genetic drift? This last is a highly controversial and difficult subject.

## REFERENCES

1. Beutler, E., and Yoshida, A.: Human glucose-6-phosphate dehydrogenase variants: a supple-supplementary tabulation. Ann Hum Genet 37: 151, 1973.

2. Bulletin of the International Hemoglobin Center, December 1976, Augusta, Georgia 30902, U.S.A.

3. Harris, H.: Enzyme polymorphism in man. Proc Roy Soc B 164: 298, 1966.

4. Harris, H., Hopkinson, D.A., and Robson, E.B.: The incidence of rare alleles determining electrophoretic variants: data on 43 enzyme loci in man. Ann Hum Genet 37: 237, 1974.

5. Harris, H., and Hopkinson, D.A.: Handbook of Enzyme Electrophoresis in Human Genetics, North-Holland Publishing Company, Amsterdam, 1976.

6. Hopkinson, D.A., and Harris, H.: Rare phosphoglucomutase phenotypes. Ann Hum Genet 30: 167-181, 1966.

7. Levy, H.L.: Genetic Screening. Advances in Human Genetics. Harris, H., and Hirschhorn, K. (Eds.), Plenum Press, New York, 1973.

8. Lewontin, R.C., and Hubby, J.L.: A molecular approach to the study of genetic heterozygosity in natural populations. II. Amount of variation and degree of heterozygosity in natural populations of Drosophila pseudoobscura. Genetics 54: 609, 1966.

9. Neel, J.V., Ueda, N., Satoh, C., Ferrell, R.E., Tanis, R.J., and Hamilton, H.B.: The frequency in Japanese of genetic variants of 22 proteins. V. Summary and comparison with data on Caucasians from the British Isles. Ann Hum Genet (in press).

10. Yoshida, A., Beutler, B., and Motulsky, A.C.: Table of human glucose-6-phosphate dehydrogenase variants.
Bull Wld Hlth Org 45: 243, 1971.

# *Analysis of Human Evolutionary Processes*

Luigi L. Cavalli-Sforza

## CHANCE VERSUS NECESSITY

In 1968, Kimura (19) hypothesized that molecular evolution, as exemplified by the differences in the molecular structure of the same protein observed in different organisms, is the outcome of chance variation. In other words, mutations responsible for protein differences, most of which are due to substitutions of one amino acid with another, are unaffected by natural selection. Being "selectively neutral," these mutations may more easily spread through species and eventually be fixed by the random evolutionary processes which are usually called "random genetic drift."

This hypothesis has met with both encouragement and opposition [see, for instance (20)], but some aspects of the controversy need clarification. Today probably nobody would dispute that a large number of all mutations are deleterious. These, however, do not usually get counted in molecular evolution as they practically never have a chance of being fixed. Being deleterious to carriers, they are usually eliminated by natural selection while still very few individuals carrying them are present. If recessive, they may reach somewhat higher frequencies because recessiveness partially screens them from selection, but they will still

---

Supported by Grant NIH 5 RO1 GM20467-04 and ERDA AT(04) 326 PA#33.

remain relatively rare. It is only when at least one allele other than the one which is prevalent in a population reaches a frequency of at least a few percent that we call a population polymorphic for that gene. In general, therefore, deleterious mutations will rarely, if ever, become polymorphic in a population and, therefore, have even fewer chances of ever being fixed, that is reaching 100% proportion. Under special conditions, however, they might increase to polymorphic proportions. A well known example is that of heterozygous advantage. Other instances include those of peculiar environmental niches in which the selective disadvantage disappears or the disadvantage is extremely small, and the population is very small and, therefore, most likely to undergo strong random changes of gene frequencies. A further complication is that a very slightly deleterious mutation may behave in practice like a neutral mutation; the smaller the disadvantage and the population size, the more easily this could be expected to occur (26).

Apart from special circumstances, neutral and advantageous mutations will then be those that can become fixed in molecular evolution, and it is their relative importance which has been at issue. A recent finding has added to the evidence that neutral mutations are those that are prevalent (13). This must be viewed in the light of former knowledge, namely that different protein molecules and different parts of the same protein are subject to widely different rates of substitution in molecular evolution. Perhaps the simplest way to interpret this phenomenon is in terms of the proportion of mutations that are advantageous, neutral, or disadvantageous (10). Assuming that really neutral mutations do not exist (the "selectionist" point of view), the rapidly changing molecules or parts of molecules must be those in which advantageous changes are relatively frequent. From the "neutralist" view, according to which truly advantageous mutations are rare and substitutions observed in molecular evolution are neutral, the rapidly changing molecules or parts of molecules must be those that are subject to less stringent selection against deleterious mutants. In both hypotheses, molecules that change less are those subject to more stringent selection, because almost all mutations occurring in them are very likely to make the molecule less functional and are thus selected against.

The new finding was made possible by the study of the sequence of nucleotides in the nucleic acids (in practice, in messenger RNA). Here it is known that, because of the properties of the genetic code, substitutions in the first or second nucleotide of every triplet will usually determine amino acid substitutions in the proteins for which they code; but the third nucleotide, if substituted, is far less likely to result in an amino acid substitution. Nucleic acid changes in the third nucleotides are, therefore, mostly not observable after the genetic message has been translated into a protein. Comparing the m-RNA corresponding to histones from two different sea urchin genera, which are practically identical for amino acid sequence (histones are in general very stable and show very few evolutionary changes), it was observed that first and second position nucleotides are unchanged, but third position nucleotides change frequently (Figure 1). In fact, they change basically at the same rate as the least stable proteins or parts of proteins (10). Thus a high rate of change is found when it is permissible to think that selection is not at work. This corresponds to the expectation of the neutralist hypothesis.

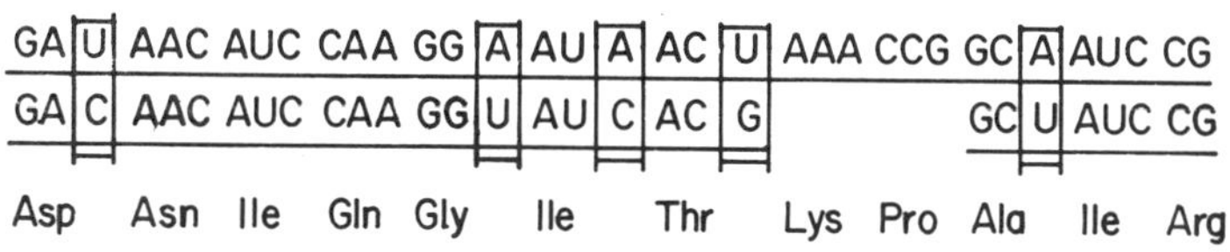


Figure 1. Part of the nucleic acid sequences corresponding to histone H4 protein of two sea urchin gerea, and the corresponding common amino acid sequence. Divergence boxed. (from M. Grunstain _et al_. 1976).

These observations need to be extended, and potential alternative explanations based on peculiar mutation rate patterns must still be ruled out; but these observations undeniably tend to support the neutralist hypothesis.

Should we then really believe that all or most polymorphic alleles are neutral, as some authors maintain? [See a summary in (25)]. It is well known, and is true also of man, that different populations of the same species may show different polymorphisms, or more commonly, different allelic frequencies of the same polymorphic genes. Only a few genes show no perceptible variation between populations of the same species living in different parts of the world. Some of this variation may well represent adaptations to local environments and a fraction of these--like sickle cell anemia --may be balanced polymorphisms. In the long run all such polymorphisms are likely to disappear, either because only one or few local races of one species will eventually survive, or because the environmental conditions favoring a balanced polymorphism disappear. Also some new genetic adaptation, not necessarily involving a balanced polymorphism, may suppress the balanced polymorphism. It is possible, therefore, that much variation between populations observed today, "synchronic" variation, reflects local variations which would not survive, were the same species examined at a later time. In other words, "diachronic" variation is not necessarily as large as the synchronic one would lead us to believe. Inferring the former from the latter may, therefore, cause an upward bias. Thus, there is certainly room for selection among existing polymorphisms, and certainly one does not expect them to be all or mostly neutral. I, for one, am tempted to abandon my former agnosticism and think that the proportion of polymorphisms which are neutral or nearly neutral may be substantial.

This statement is the same as saying, to use Monod's words (23), that "chance" is important in molecular evolution. Of course this does not deny, or greatly reduce, the role of "necessity" (in other words, of natural selection). All the "disadvantageous" mutations--all of those which are selected against almost as soon as they occur and, therefore, hardly ever become polymorphic--are, of course, examples of the action of necessity. Moreover, those amino acids which are not now easily substituted by others may have been introduced at an earlier time by "advantageous" mutations, again supporting the fundamental role of natural selection, without which, of course, no life would be possible.

Some years ago work was done to test the role of drift at another level: the variation between small populations (7,8). When enough demographic information is available, the variation due to drift is predictable. It is easier to observe drift in small populations, because here we expect drift effects to be comparatively larger. If there were local variations of selective conditions, variation of gene frequencies greater than that due to drift would be expected, while stabilizing selection (as with heterozygote advantage) might reduce it below the drift level. Actual observations carried out in villages near Parma University using standard blood markers confirmed that the genetic variation observed in the area studied could be explained primarily by drift alone. Similar results were obtained elsewhere (3). These conclusions, of course, should not be extrapolated beyond the areas for which they were reached: it is possible, indeed likely, that when larger areas--or the whole habitable earth--are examined, at least some alleles of any particular gene will show different selective values in different parts of the world.

We can reasonably infer local variation in selection for those alleles which show important differences in frequencies from one population to the other. No known human gene shows more variation than the locus controlling the constant portion of the IgG heavy chains: the Gm locus. For reasons that still escape us entirely, many alleles are found that are peculiar to some populations and reach high frequencies in them (17).

Many people, including the writer, are probably ready to believe that locally different selection is the cause of differentiation of Gm gene frequencies; but unfortunately thus far there is no fully satisfactory proof that selection is involved or of the way in which selection acts. Other human genes showing profound local differentiation are Duffy and Rh. For the first there is some evidence of selective variation, the Duffy gene (Fy) controlling synthesis of a receptor for a malarial parasite (22). It is, therefore, not surprising that an allele (Fy) causing the loss of the receptor in red cells is frequent among Africans, who have considerable exposure to the disease. One may ask why, then, for sickle cell anemia (S), thalassemia (Th), and G6PD, for all of which the selective role of malaria is well established, there is no variation of incidence as striking as that observed for Fy or Rh. For S and Th this is easily understood, given that fixation of the allele conferring resistance to malaria cannot be achieved, since S or Th are almost lethal when homozygous (17). For G6PD there is at least one allele, however, and possibly more, which are perhaps not so deleterious, as shown by the fact that they have reached very high frequencies (60-65%) in some populations (24). Maybe these alleles are or were on their way to fixation in these populations; now, however, that malaria is being wiped out, the course of evolution will very probably be altered.

## RECONSTRUCTION OF EVOLUTIONARY HISTORY

Today it is clearly conceivable that a substantial fraction of the interracial variation for genes other than those just cited may be due to drift. This is of some interest in reconstructing evolutionary history, a task which was attempted with surprisingly good results in the early sixties using gene frequency data (19). Populations known to be geographically or historically closely related were also found to be closely similar when genetic distance was evaluated, using cumulated information obtained from a relatively small number of genes. Trees of origin were thus constructed using the principle that genetic distance increases monotonically with the time of separation of two populations.

Monotonicity of genetic distance with separation time--a prerequisite for validity of the analysis--is expected to hold, when enough genetic markers independently evolving under random genetic drift are considered simultaneously. But monotonicity need not apply if selection is involved, and in such a case the reconstruction of evolutionary history may be wrong.

Basically the same techniques of evolutionary history reconstruction that were used for human races were also employed sucessfully for the analysis of molecular evolution. This analysis is based on the comparison of the structure of proteins in different animals (and also plants and microorganisms). In evolution, during long intervals of time which separate two species of animals--for instance, a mammal and a bird (on the order of 200 million years)--many allelic fixations have occurred and a given protein is found to differ in the two species by several amino acid substitutions. The number of substitutions is then a measure of genetic distance, which is proportional to the time of evolutionary separation; but human races have differentiated for intervals of time which are well below a million, and probably below 100,000 years so that differences between races involving the fixation of different alleles are obviously expected to be rare, and indeed are highly unusual. Only a small fraction of a fixation has had the time to occur, on average, where races of a species are compared, and that is why genetic distances between human races are reduced to differences in gene frequencies. The overall magnitude of genetic distance is thus much larger in molecular evolution, in which, traditionally, widely different species are compared than in human interracial comparison. In addition, in interspecific evolution (as in molecular evolution) migration between species is excluded, while it is present as a potential cause of complication in interracial evolution.

The success of phylogenetic reconstructions in molecular evolution is understood more easily once it is accepted that much of molecular evolution is due to random genetic drift. It should be added, however, that a type of hybrid between selection and drift, selective drift, can also help in explaining why evolutionary reconstruction can be satisfactory even if random genetic drift is not the only cause of divergence. To assure monotonicity of the relation between time and genetic distance, randomness of the source of divergence is basic. It alone assures the validity of the expectation, that genetic distance (average over enough genes) increases monotonicity with time but randomness may be generated also by random changes of selection coefficients over time and place. This is the definition of selective drift (18) which, like random genetic drift, can also assure the monotonicity of the relationship necessary for the validity of evolutionary reconstructions. The principle for validity of phylogenetic analysis is, therefore, that for reconstructing evolutionary history, monotonicity of genetic distance between two populations and time since their separation is required. This demands that the process of evolutionary differentiation be basically random, that is, largely independent of the times and places at which the evolutionary changes take place. Another problem, that of variation of evolutionary rates, will not be discussed here.

It is instructive to consider a case in which this principle was probably violated. When Edwards and I made the first reconstruction of human evolution, based on gene frequencies at the five loci which showed maximum discrimination among those on which information was then available (ABO, Rh, MNS, Di, Fy), we also tried to use independent anthropometric measurements [including skin color (9)]. The results were very similar, except for a very important detail: the first split. With both types of data, related populations grouped together: Africans with Africans, Europeans with Europeans, and so also for American natives, and Orientals; but the four population groups formed different clusters when considering genes and when considering anthropometric markers (Figure 2). The study of further genetic markers confirmed the original classification with five blood group loci (6,16,11, 27). The analysis of skull measurements of 17 ethnic groups carried out recently by Howells confirmed in full the anthropometric type of clustering (14,15); the paradox, being confirmed, required an explanation. I think today we can give that explanation (21). Anthropometrics (including skin color) measure mostly external traits, and these are very sensitive to climate to which they show both short term (physiological) and long term (genetic) adaptation. It is not surprising, therefore, that the first split, as obtained on the basis of anthropometric data, is basically a split between climatic zones. It does not, necessarily, represent evolutionary history; but genes are far less sensitive to climate.

The 70 skull measurements were condensed by Howells using discriminant analysis into fewer functions. The first discriminant function accounted for over 25% of all the variations and

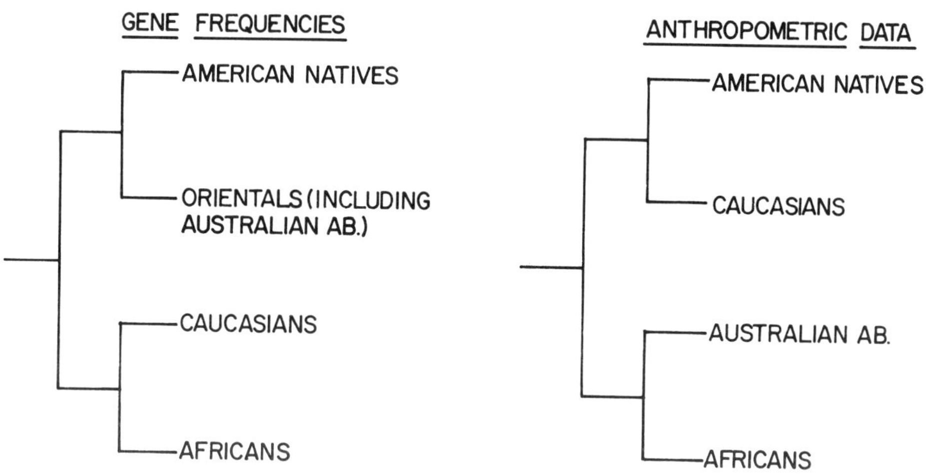


Figure 2. Grouping of population clusters in evolutionary trees based on gene frequencies and anthropometric data.

the second, for about 15%. The first discriminant separates populations according to the anthropometric tree and the second, to the genetic tree. [For the relationships of trees and vectorial representation as in discriminant analysis, see (11)]. Table 1 shows that the first discriminant function is highly correlated with climate, in particular temperatures, while the second and practically all other discriminant functions are not. It may also be remembered that there is important evidence from classical sources--in organisms other than man--that external measurements show climatic variation, which is very probably adaptive.

Thus, anthropometrics may be so sensitive to an important environmental variable--climate--that

Table 1

Correlation of Howells' first two discriminant functions for skull measurements with climatic variables (from R. Matessi *et al.*)

| Climatic Variables | Function 1 | | Function 2 | |
|---|---|---|---|---|
| | Males | Females | Males | Females |
| Average temp. in a year | -.78*** | -.72*** | .04 | 23 |
| Abs. max temp. | -.51* | -.40 | .15 | 10 |
| Abs. min. temp. | -.77*** | -.76*** | -.19 | -.21 |
| Mean relative humidity | .46 | .41 | -.17 | -.21 |
| Rainfall in a year | -.22 | -.21 | -.13 | .04 |

* significant at 5%
*** significant at 0.1%

they cannot be trusted as indicators of evolutionary history. The perfect indicators, as we have discussed, are those showing fully random variation. As far as we are aware today, genes come closer to this requirement and, therefore, genetic markers are likely to be more trustworthy indicators of evolutionary history than anthropometrics.

## THE ORIGIN OF CAUCASIANS

Caucasians offer some interesting peculiarities from an evolutionary point of view which can be explained on the basis of their history. Starting with the age of the great geographic discoveries (XV century), Caucasians, under demographic pressure, colonized the rediscovered continents which were at the time relatively underpopulated: the Americas and Australia. Even before that time Caucasians already occupied a very large territory, shaped like a rough ellipsoid with extremes in Iceland and India, spanning a continent and a half. In spite of the wide territory occupied, they were fairly homogeneous. We can get an idea of the genetic variation between Caucasians, as compared with that in other population clusters, by looking at an analysis by principal components of world gene frequency data (based on 58 independent alleles). Figure 3 is such a map [from work by Piazza *et al*. (27)]. Some Caucasian groups, like Indians (the population represented in the figure is from East India) depart somewhat from the cluster (in the expected direction, e.g., toward Orientals with whom they are likely to have had some admixture); but most Caucasians cluster closely together, especially if compared with natives of America or Oceania.

In a representation of the same data by trees, Caucasians attach to the tree in a very peculiar way: they join on by short branches near the root of it (11,27). This would be expected if admixtures of earlier groups had an important part in their origin. To understand this, consider that

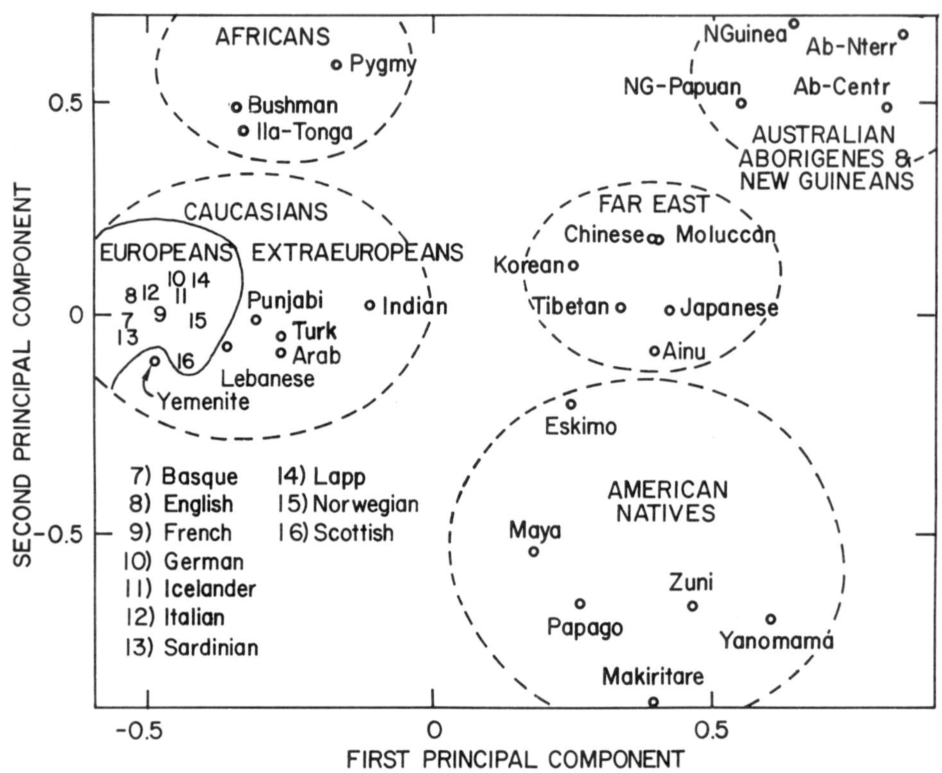


Figure 3. Analysis using data from Evian workshop (58 independent markers including HLA).

our simplest guess about the more remote ancestors of modern man (at the root of the tree) is obtained by taking the average gene frequencies of presently existing groups. (This, of course, ignores possible selective trends in gene frequencies which may have affected, in parallel, all human populations; but such trends could hardly ever be detected on the basis of comparisons between living groups). If the best guess of the gene frequencies of our ancestors is given by the average of gene frequencies found today, a population obtained by hybridization between presently living groups would obviously imitate such ancestral populations and be located, in the tree, near the root to which it would be attached by a branch of length zero or close to it.

Although the actual situation observed for Caucasians is not so extreme, it is a fact that both they, and some Oriental populations (in particular Chinese) come very close to it. The explanation of this phenomenon is probably to be found in one important reason for admixture: at the center of the area inhabited by Caucasians (and also, in China and Southeast Asia) we find the beginnings of plant and animal breeding. Concentrating on Caucasians, for whom more is known both from an archaeological and from a genetic point of view, it has been ascertained that the domestication of many cereals and also of animals such as cattle, pigs, sheep, and goats took place in the Middle East in a period about 10,000 to 9,000 years ago. From there the farming economy spread slowly, but at a regular rate in all directions. The spread toward Europe is far better known than that in other directions, given the more extensive archaeological knowledge available in Europe. Ammerman and I have mapped the progress of agriculture in Europe on the basis of existing archaeological records and radiocarbon (1). This progress was found to take place fairly regularly at an average rate of one Km per year. Thus, it took, for instance, 3,000 years for agriculture to diffuse over the approximately 3,000 kilometers separating the Middle East from England.

Diffusion of agriculture may have taken place in two ways (2) and probably both occurred but their relative importance has to be assessed: 1) Farmers themselves may have spread (demic diffusion). This would be expected if the transition to farming were accompanied by population growth. A farming economy is very likely to do that, a) by increasing the carrying capacity of the land over that prevailing in previous economies (hunting/gathering), b) by increasing the motivation for breeding larger families (hunters/gatherers usually have smaller familes than farmers), and c) by decreasing mortality with a decrease in the likelihood of famine (food being more easily stored). 2) Stimulus or cultural diffusion is also possible: hunters/gatherers may have acquired the new agricultural techniques and life-style from neighboring farmers.

The diffusion of agriculture was sufficiently slow that demic diffusion may have played an important role. Too fast a diffusion process would have been incompatible with demic diffusion; but a slow process, of course, does not of itself exclude cultural diffusion. Genetic analysis may help in answering the question, for if the diffusion of agriculture were entirely demic, Caucasians should be perfectly homogeneous genetically; if it were entirely cultural, it would not have affected the preexisting genetic pattern, which was likely to be highly variable geographically (as is true of all other situations known in the world in which agriculture did not develop or did not involve extensive population growth). If both mechanisms participated, one should find a pattern of diffusion of genes with a center located in the Middle East and great variation at the periphery. Naturally, we must assume that later events have not altered too markedly gene frequencies: i.e., that latter diffusion from the outside, internal immigration, or later selection processes have not been of such a magnitude as to destroy the pattern created by the diffusion of agriculture.

The overall picture indicates that both demic and cultural diffusion took place perhaps with an overall predominance of the demic over the cultural component. Individual genes which show substantial variation over the area involved agree with this conclusion. The best indicator is perhaps the Rh gene, for which there exists a very substantial geographic variation over the whole area [see maps in (24,4)]. A reasonable hypothesis is that $R_1$ and $R_2$ alleles were the alleles of highest frequency in the Near East (and also further to the East) before the spread of agriculture, while hunters/gatherers of western Europe were perhaps 100% Rh negative and those of Africa 100% $Rh_0$.

This would explain the gradient of decreasing frequency of Rh positive genes (of $R_1$ and $R_2$ type) from the Middle East toward Europe, as well as toward Africa, and also those of decreasing Rh negativity from western to eastern Europe and of the $Rh_0$ allele from central Africa to northern Africa and Asia. In the direction of India the gradients are not so sharp but this is not surprising.

This explanation is in agreement with a suggestion advanced by Mourant some time ago [discussed in (24)] that the Rh negative gene was characteristic of early Europeans. This hypothesis was derived from the observation that Basques (living near the Pyrenees at the boundary between France and Spain) have a high frequency of Rh negative genes, around 55%. It may be added that figures of Rh- frequencies around 50% are especially puzzling. Selection in Rh incompatibility is directed against heterozygotes, and Rh gene frequencies, therefore, tend away from the 50% region. Those gene frequencies that are even slightly above 50% tend fairly rapidly to 100%, and those below 50% tend to 0% almost as rapidly. What, if anything, maintained these frequencies around 50%? The simplest explanation in agreement with what we have seen is that changes of gene frequencies determined by incompatibility selection were slowed down by migration.

The full story, then, may be as follows: Very high Rh- gene frequencies, perhaps close to 100% probably formed among western and northern Europe hunters/gatherers prior to the arrival of the farmers. Apart perhaps from rare periods of higher population densities, these hunters were probably at very low densities most of the time. The whole population of English hunters has been estimated at some 10,000 individuals (12). Moreover, these low densities have lasted over long periods. Under these conditions, drift may have made it possible to establish a population that was almost entirely Rh negative. Drift effects, it may be noted, are proportional to t/N, where _t_ is time in generations, _N_ population size. Even if _N_ is large, there may still be substantial drift effects, if _t_ is also large. Following the spread of farmers, R̄h positive genes were then introduced (or reintroduced) into Europe by farmers. The mechanism of selection for incompatibility, which eliminates Rh positive genes when Rh negative genes are numerically prevalent, must have helped keep the Rh negative gene frequency detectably high among descendants of hunters/ gatherers in western and northern Europe, and must have been fighting successfully the incoming Rh positive genes for the immigrant farmers, as Basques have still somewhat more than 50% Rh negative genes.

It is likely that the evolution of Gm genes is too fast to be trusted as a historical probe of this migratory phenomenon that started some 10,000 years ago. Moreover, gene data available are not sufficient or entirely satisfactory. It is very encouraging that HLA genes are found to be in basic agreement with the hypothesis (29).

All other genes are less informative than these, but one can hope for obtaining a clearer picture by correlating the information from many genes. Attempts at doing this (with Sgaramella-Zonta) by a method previously published (28) are in agreement with the idea that demic diffusion has been of great importance in the spread of agriculture. Attempts at estimation of the relative roles of demic and cultural diffusion are now taking place. Thus it seems reasonable to suggest that Caucasians are, to a large extent, descendants of Middle East farmers, who mixed with a variety of preagricultural populations they found in the areas into which they migrated. Most probably the original farmers, being at the crossroads of three continents, were themselves a complex admixture.

## JEWISH DIASPORAS FROM A GENETIC POINT OF VIEW

An especially interesting example of genetic variation is found among the Jews. Twice (in 500 B.C. and, more extensively in 70 A.D.) Jews have left Israel in great numbers and formed colonies outside the area of origin,in many different parts of the world. Most of the colonies thus formed have been relatively isolated genetically from neighboring people and have preserved their culture more or less unchanged for 1,900 years or more. As is well known, many such groups have recently returned to Israel. Here genetic testing on these groups has taken place, much of it due to the efforts of Israeli workers.

Carmelli *et al*. (5) have carried out an analysis of genetic markers. Three major conclusions can be reached from this analysis: 1) there is no distinct "Jewish race": the Jewish populations scatter over the entire genetic map more than nonJewish ones. 2) Even so, most Jewish groups show more similarity with people from the area of their remote geographic origin (the Middle East) than with those people with whom they were associated in the last 2,000 years. 3) Even though Jewish populations on average still resemble Middle East populations, so that the estimates of their original gene frequencies fall near the center of the Near East cluster, their gene frequencies have scattered around their origin a great deal, probably due to drift. Estimates of population effective sizes derived from this scatter vary from over 1,000 to more than 10,000, or 50,000, depending on the population, and are compatible with historical data. Further historical research may help in providing direct estimates of these quantities, which would be of importance in refining the analysis.

## CONCLUSIONS

Human evolution shows an intricate interplay of selection and drift. In favorable circumstances the relative role of the two can be dissected. Drift plays a role not only at a microgeographic level and in highly isolated populations, where it is relatively more important, but also in determining variation over wide geographic areas, either earlier in human history, when population densities were very low, or because drift effects in large populations became noticeable because they acted over a long time. The increasing evidence of the importance of drift in molecular evolution supports this interpretation but it does not lessen the importance of selection.

Under special conditions, of which the agricultural revolution was an outstanding example, favoring the growth and spread of Caucasians (and of Chinese), there may be extensive hybridizations which tend to cancel previous diversifications. The relative homogeneity and superficial lack of evolution among Caucasians are consequences of such growth and spread sparked by important technological innovations in food production. Interactions of cultural developments and biological evolutions have been especially frequent in the evolutionary history of man. Another important example is provided by the diasporas of the Jews.

The kinds of reconstruction of evolutionary history which we have attempted here are, of course, difficult and supportive evidence is generally weak. Accumulation of evidence from different sources is the major way which should eventually permit us to put a reasonable amount of confidence in our conclusions.

## SUMMARY

Most recent trends in the interpretation of molecular evolution indicate that many mutations are deleterious, the rest largely neutral, with few advantageous changes. This encourages us to believe that a non-negligible fraction of polymorphisms found in races of one species may be neutral. In reconstruction of phylogenetic analysis it is important to use, as much as possible, traits subject to random variation; neutral mutations are, therefore, ideal. This helps in explaining the success in reconstruction of phylogeny from molecular evolution data. The same reconstruction in human differentiation seems to be more successful with genes than with anthropometric traits, the latter being more frequently subject to adaptation to local conditions. The effect of climate was shown to be important in making one set of anthropometric data disagree with phylogenetic analysis based on genetic markers.

The differentiation among Caucasians is probably the consequence of the spread, from a center of origin in the Middle East, of farmers who migrated slowly in all directions into a variety of areas which were already inhabited by hunters/gatherers who were probably highly differentiated and with whom farmers mixed to varying degrees.

Another example of variation is that of Jews in the two diasporas. As most Jewish colonies were small, at least for part of the time, substantial drift is observed. Admixture with local populations is mostly limited.

REFERENCES

1. Ammerman, A., and Cavalli-Sforza, L.: Measuring the rate of spread of early farming in Europe. Man 6: 674-688, 1971.

2. Ammerman, A., and Cavalli-Sforza, L.: A population model for the diffusion of early farming in Europe. In The Explanation of Culture Change, C. Renfrew (Ed.), G. Duckworth & Company, London, 1973.

3. Bodmer, W., and Cavalli-Sforza, L.: The analysis of genetic variation using migration matrices. In Genetic Distance, J.F. Crow and C. Denniston (Eds.), Plenum Publishing Company, New York, 1975, pp. 45-61.

4. Bodmer, W., and Cavalli-Sforza, L.: Genetics Evolution and Man, W.H. Freeman Company, San Francisco, 1976.

5. Carmelli, D., and Cavalli-Sforza, L.: The genetic origin of the Jews: A multivariate approach. (ms being submitted).

6. Cavalli-Sforza, L.: Population structure and human evolution. Proceedings Royal Society of London 164: 362-379, 1966.

7. Cavalli-Sforza, L.: Genetic drift in an Italian population. Scientific American 221(2): 30-37, 1969.

8. Cavalli-Sforza, L.: Some current problems of human population genetics. Am J Hum Gen 25: 82-104, 1973.

9. Cavalli-Sforza, L., Barrai, I., and Edwards, A.W.F.: Analysis of human evolution. Genetics Today (Proceedings of the XI International Congress of Genetics, Vol. 2), The Hague, Netherlands, 1963, pp. 932-952.

10. Cavalli-Sforza, L.L., and Bodmer, W.: The Genetics of Human Populations, W.H. Freeman Company, San Francisco, 1971.

11. Cavalli-Sforza, L., and Piazza, A.: Analysis of evolution: evolutionary rates, independence and treeness. Theor Pop Biol 8: 127-165, 1975.

12. Clark, J.Y.D.: Star Carr: a case study. Addison Wesley Modules in Anthropology, No. 10, Addison Wesley, Menlo Park, California, 1972.

13. Grunstein, M., Schedl, P., and Kedes, L.: Sequence analysis and evolution of sea urchins (Lytechinus pictus and Strongylocentrotus purpuratus) Histone H4 Messenger RNAs. J Mol Biol 104: 351-369, 1976.

14. Howells, W.: Cranial Variation in Man. Papers of the Peabody Museum of Archeology and Ethnology, Vol. 67, Harvard University, 1973.

15. Howells, W.: Explaining modern man: evolutionists versus migrationists. J Hum Evol 5: 477-495, 1976.

16. Kidd, K.: Genetic approaches to human evolution. Atti del Colloquio Sul Tema: L'origin dell'Uomo Accademia Nationale de Lincei Rome, 1973, pp. 148-174.

17. Kimura, M.: personal communication.

18. Kimura, M.: Process leading to quasifixation of genes in natural populations due to random fluctuation of selection intensities. Genetics 39: 280-295, 1954.

19. Kimura, M.: Evoutionary rate at the molecular level. Nature 217: 624-626, 1968.

20. LeCam, L.M., Neymon, J., and Scott, E.L. (Eds.): Proceedings of the Sixth Berkeley Symposium on Mathematical Statistics and Probability, University of California Press, Berkeley, California 1971.

21. Matessi C. et al.: Conditions for the evolution of altruism under Darwinian selection. Theor Pop Biol 9: 360-387, 1976.

22. Miller, L.H. et al.: Resistance factors to Plasmodium Vivax in Blacks: The Duffy-blood-group genotype, FyFy. New Engl J Med 295: 302, 1976.

23. Monod, J.: Le Hasard Et La Nécessité. Editions Du Seuil, Paris, France, 1970.

24. Mourant, A.E., Kopec, A.C., and Domaniewska-Sobczak, K.: The Distribution of the Human Blood Groups and Other Polymorphisms, Oxford University Press, London, 1976.

25. Nei, M.: Molecular Population Genetics and Evolution, North-Holland Publishing Company, Inc., New York, 1975.

26. Ohta, T.: Slightly deleterious mutant substitutions in evolution. Nature 246: 96-98, 1973.

27. Piazza, A., Sgaramella-Zonta, L., Gluckman, P., and Cavalli-Sforza, L.: The Fifth Histocompatibility Workshop. Gene Frequency data: A phylogenetic analysis. Tissue Antigens 5: 445-463, 1975.

28. Sgaramella-Zonta, L., and Cavalli-Sforza, L.: A method of the detection of a demic cline. Proceedings, Workshop on Population Structure, Hawaii University Press, Honolulu, Hawaii, 973, pp. 128-135.

29. Thomson,, G., Bodmer, W., and Bodmer, J.: The HLA system as model for studying the interaction between selection migration and linkage. (in press), 1976.

# HLA Polymorphism and Disease Associations

Italo Barraï

Pier Luigi Mattiuz

The HLA system, or human histocompatibility system, occupies a complex region located on chromosome 6 of man. The region comprises a chromosome segment approximately 18 to 20 thousandths of a recombinational unit, or close to two centimorgans, in length. It was suggested by Bodmer (3) that a region of such length might include a minimum of 200 genes, even under the assumption that only ten percent of the chromosomal DNA is active. Should all of the DNA be active, the HLA region might encompass more than 2000 genes. Four main loci have been identified in the region, all closely linked; they are called A, B, C, and D, respectively.

Since the discovery of the importance of the antigen specificities determined by the region in tissue compatibility, an enormous amount of work has been invested in this system, initially in view of the relevance it had to procedures of organ transplantation, particularly of kidneys. The most immediate and striking finding was the complexity of the region. It was found that each locus in it is characterized by several alleles which are present with different frequencies in different populations; and population studies, as well as family studies, have demonstrated the extreme degree of existing and potential polymorphism of this system.

---

With the partial support of the World Health Organization to the Institute of Zoology of the University of Ferrara.

## POPULATION CONSIDERATIONS

The loci of the HLA system for which there is the most information are the HLA-A and HLA-B. In Caucasian populations of Europe, the Middle East and Asia, locus A is characterized by 12 alleles; in the same populations, the average number of alleles at locus B is 15. No significant difference was found in the number of alleles present in different Caucasian populations for the two loci A and B, which are also called "serum defined" from the technique through which their relative antigenic specificities are detected. In Mongolian populations, the average number of alleles is 10 and 12 respectively for the two loci. Populations of Oceania have an average of 5 and 8 for A and B respectively, and Australian populations have corresponding averages of 8 and 5. Negroid populations from Africa have average numbers of alleles at the two loci of 12 and 14. The results from population studies are summarized in Table 1, in which the expected number of haplotypes corresponding to the two loci is also given.

Table 1. Number of HLA-A and HLA-B alleles in six human groups.

| Population | Number of samples | Allele HLA-A | Allele HLA-B | Number of haplotypes |
|---|---|---|---|---|
| Caucasians | 17 | 12 ± 1 | 15 ± 1 | 180 ± 26 |
| Mongoloids | 9 | 10 ± 2 | 12 ± 1 | 118 ± 36 |
| Oceanians | 4 | 5 ± 1 | 8 ± 2 | 39 ± 18 |
| Australians | 2 | 8 ± 3 | 5 ± 0 | 40 ± 14 |
| Amerindians | 8 | 7 ± 1 | 9 ± 3 | 67 ± 24 |
| Negroids | 6 | 12 ± 1 | 14 ± 1 | 161 ± 14 |
| Averages and totals | 46 | 10 + 3 | 12 + 3 | 128 + 59 |

The overall data show that on the average, in man, the locus A has 10 alleles, and the locus B, 12 alleles, each with considerable frequency. The expected average number of haplotypes, considering only the two loci, A and B, is 128, from which the expected number of genotypes is more than eight thousand; the possible phenotypes are more than four thousand. More recent data of Bodmer and Thomson (7) show that the potential variation is much higher for the whole system. Considering all the four loci, A, B, C, and D, and their alleles, of which 19, 26, 6, and 9 respectively have been defined, the possible number of haplotypes is more than 27,000; the possible number of genotypes is above 350 million, and the possible phenotypes are more than 33 million.

The average heterozygosity is 88% for A, 92% for B, 44% for C, and 76% for D. It was calculated separately for each locus. Since the loci are closely linked, and there is often linkage disequilibrium between alleles at different loci, it is not possible to compute exactly the expected heterozygosity at any one of the four loci; however, it is apparent that heterozygosity at one or more of the four loci has a very high probability, close to 100%. With such a high number of alleles, the system maintains almost forced heterozygosity for at least one of the loci. Homozygotes have been observed in the progeny of consanguineous marriages (2).

Two considerations which can be made immediately, in the presence of such a polymorphism are 1) how the polymorphism was generated, and 2) how it can be maintained. We prefer to discuss first the mechanisms which could possibly maintain such an enormous variability, once it is generated; and secondly, the possible mechanisms which have generated the variability itself. We prefer this order of discussion because it is not apparent, at the present time, whether there is a strong selective factor responsible for the maintenance of the polymorphism, say one mechanism as convincing as the one which seems to operate for some hemoglobin variants in malarious environments. Therefore, one might expect to observe the effect of genetic drift on the system, since the HLA polymorphism seems to be composed of a multitude of variants, each one with a small advantage, or with no appreciable advantage, and also with some disadvantage, as in the case of disease susceptibility.

## THE EFFECT OF RANDOM GENETIC DRIFT

The classical theory of random genetic drift, as examined and described by Kimura (9) states that, everything else being fixed, the probability of coexistence of alleles present at a given time at one locus in one population, decreases rapidly with the number of alleles. In fact, the rate of exponential decrease in time is proportional to the square of the number of existing alleles. In the case of the HLA locus, each haplotype behaves as a complex allele, and drift would greatly decrease the number of haplotypes in a few generations, under the assumption that the population size is large with respect to the number of haplotypes. Since this assumption is one of the main restrictions of the theory, it seems that it is not possible to apply the classical theory of random genetic drift to the HLA system, because the number of haplotypes is of the same order of magnitude as the effective breeding size.

Once the extreme variability at the HLA locus is realized, and given the breeding patterns of humans, in whose groups there is almost invariably some avoidance of close inbreeding, it seems unlikely that the classical rules of decay of variability apply for this system. This statement is based on the requirement of the classical theory, which needs the number of alleles to be small in comparison to the size of the breeding unit in order that the rate of decay of variability be valid. It would follow that the enormous variability existing at the system does not require a particular mechanism to be maintained. Once it is in a population, it stays there, because the rules of decay of the variability do not hold anymore.

One may well visualize a limiting theoretical case, an infinite population with an infinite number of alleles, where the number of alleles stays infinite in absence of any selection.

The assumption of selection is then not strictly necessary for the maintenance of the polymorphism at its present degree of variability. However, it seems unavoidable to postulate selection as a causative agent for the generation of the polymorphism. Namely, even if selection were no longer necessary for the maintenance of the polymorphism now, it must have acted in the past to generate the diversification. It is also true that differential mutation rates might have generated that situation we are discussing for HLA. In fact, there are proponents of the hypothesis of differential mutation rates at the locus.

Once we have accepted that new HLA variants arise because of mutation at the allele level, and because of recombination at the haplotype level, we want to determine whether differential mutation rates are likely to have generated the polymorphism in its present frequencies.

## THE MUTATIONAL HYPOTHESIS

The absence of natural selection, mutation and backmutation can maintain a polymorphism at any level, as long as the rate of mutation counteracts the effect of subdivision, isolation by distance, and genetic drift. Since the variation at the HLA locus is so high, one would therefore expect that there is a correlation with an increased mutation rate at this locus. The mutational hypothesis, however, is difficult to defend for several reasons. The first is that there is no evidence, in man, for differential mutation rates. This might be due to insensitivity of our present methods for estimating mutation rates in man; in fact, it was observed that mutation rates estimated at the same locus in man vary on the average by a factor of three (1). Such variation, which might be due to methodological procedures and/or sampling accidents, makes it difficult to entertain a strong opinion on the question of differential rates of mutation (Table 2). As a rule, mutation is a property of DNA and not of the locus, and, given the anatomy of DNA, it would be expected that mutation rates would be constant for loci of the same length. In the case of the HLA region it would not be possible to entertain the opinion of a higher mutation rate based on the observed number of alleles at the same locus, unless even the locus is a complex entity of great length. Even in man, a large number of alleles exist at other loci, but with frequencies which are very low. For example, one may cite the case of the hemoglobin variants, which are very numerous, but with an extremely low frequency except for a few alleles which are favored by selection. In the case of HLA, the distance between loci would allow the formation of new haplotypes through recombination, with a frequency which is at least two orders of magnitude larger than the average mutation rate in man.

Table 2. Variability of estimates of mutation rates at the same locus in man.

| TRAIT | MUTATION PER MILLION GAMETES | | |
|---|---|---|---|
| | Minimum (m) | Maximum (M) | Ratio M/n |
| Diaphyseal aclasis | 6 | 9 | 1.5 |
| Epiloi | 8 | 16 | 2.0 |
| Marfan's syndrome | 4.2 | 5.8 | 1.4 |
| Osteogenesis imperfecta | 7 | 13 | 1.8 |
| Retinoblastoma | 7 | 46 | 6.6 |
| Duchenne muscular dystrophy | 43 | 95 | 2.2 |
| Facioscapulohumeral muscular dystrophy | 0.5 | 5 | 10.0 |
| Haemophilia A | 13 | 57 | 4.4 |
| Hypogammaglobulinemia | 3 | 5 | 1.7 |
| Achondroplasia | 10 | 13 | 1.3 |
| Polycystic disease of the kidney | 65 | 120 | 1.8 |
| AVERAGE | 15 | 35 | 3.2 |

## THE SELECTION HYPOTHESIS

According to R.A. Fisher, natural selection is a process which leads to a high degree of improbability. The extreme variability of the HLA system, although not unique in nature (for example, besides the H2 locus in mouse, there are systems highly polymorphic in serum groups in cattle) is an event which cannot easily be explained as a result of neutral processes; therefore, the alternative which emerges strongly is the one of selection. However, at the present time, there is no strong evidence of selection of such a type which might justify this extreme polymorphism. No differential viability, fertility, or feto-maternal incompatibility have been measured for the alleles or the haplotypes of the HLA locus (10). Indirect evidence for the presence of selection was given by Bodmer, Cann, and Piazza (5). There are, however, observations which might be a clue to the solution of the problem or to the formulation of an acceptable hypothesis. At the 1972 HLA Workshop it was reported that certain groups, in particular the American Indians and the islanders of Oceania, have a smaller number of alleles and a higher degree of homozygosis than the other populations. Since this smaller number is characteristic only of their HLA system, and not of the other systems, Bodmer (4) inferred that the difference could not be due to genetic drift, which should affect equally all loci, but must be due to natural selection. For American Indians, selection might have been represented by infectious or parasitic diseases which were typical of the continent at the time of colonization, and which wiped out those HLA alleles which were not associated with resistance to specific diseases. A phenomenon which would further speak in favor of disease resistance (or sensitivity) for the HLA locus is the fact that when the same populations again came in contact with diseases to which they had not been exposed for many generations, they were sensitive again to the diseases, possibly because some alleles for resistance had been lost through genetic drift in the period during which the populations were isolated; the loss may have occurred since these alleles were no longer advantageous in the new environment. Population bottlenecks and founder effects at the time of migration might simulate drift effect.

Direct association between HLA and disease has been the object of much study recently. About 20 diseases have been found to be associated with one or more HLA antigens (Table 3). The strength of the association is variable [see for analysis of statistical associations Svejgaard *et al.*(12)]; in some cases, as in ankylosing spondylitis, it is extreme, close to one hundred percent. In other diseases it is weak, as in multiple sclerosis and acute lymphatic leukemia. However, the evidence for association with disease is strong. It was noted that most of the diseases which have been found to be associated with HLA antigens are rare, so that it seems unlikely that the diseases which are found to be associated at present represent, or ever represented, a strong selective factor (11). The indication that is obtained by the observed associations is that there is the possibility that other antigens might have been associated with sensitivity to more widespread and selective diseases, such as influenza, smallpox, and others causing severe mortality; and that the present alleles are extant because they are resistant to the present infections which could otherwise be lethal; in fact, they could be the result of selection for resistance to diseases of this type.

The hypothesis that HLA might be responsible for resistance to pathogens is now advocated by several authors, notably by Bodmer's group. Under this hypothesis, new variants would be advantageous, since the resistance to pathogens implies that a pathogen has to adapt to the antigenic or immune response status of the host in order to be successfully virulent. Any new variant in HLA would disrupt the previous status of adaptation of the pathogen and would confer some additional resistance to that pathogen. In other words, the mechanism would show some degree of analogy with the phenomenon of insecticide or drug resistance

Table 3. Association of HLA with disease.

| Disease | Antigen | Risk | Reference |
|---|---|---|---|
| Ankylosing spondylitis | HLA-B27 | 121-141 | 11,12,8 |
| Reiters's disease | HLA-B27 | 4-47 | 11,12 |
| Acute anterior uveitis | HLA-B27 | 17-30 | 11,12 |
| Psoriasis (several types) | HLA-B13<br>HLA-Bw16<br>HLA-Bw17 | 5 | 11,12 |
| Graves' disease | HLA-B8 | 3-4 | 11,12 |
| Coeliac disease | HLA-B8 | 10 | 11,12 |
| Myasthenia gravis | HLA-B8 | 4-5 | 11,12 |
| Chronic hepatitis | HLA-B8 | 4-10 | 11,12 |
| Hodgkin's disease | HLA-B8<br>HLA-Bw35<br>HLA-A1 | 1,3 | 11 |
| Acute lymphatic leukemia | HLA-A2 | 2 | 11 |
| Systemic lupus erythematosus | HLA-Bw15 | 5 | 11 |
| Burkitt's lymphoma | several | 1 | 6 |
| Multiple sclerosis | HLA-A3 | 2 | 11 |
| Haemophilus influenza | HLA-Bw17 | 4 | 12 |
| Infectious mononucleosis | HLA-Bw35 | 3 | 12 |
| Chronic glomerular nephritis | HLA-A2 | 1,5 | 12 |
| Juvenile diabetes mellitus | several | 3 | 12 |

6. Bodmer *et al.*, 1975
8. Hill *et al.*, 1976
11. McDevitt and Bodmer, 1974
12. Svejaard *et al.*, 1975

in several organisms. This observation is based on the fact that populations which have long been separated from a pathogen--for example, Amerindians from Caucasian pathogens--may have lost the alleles and the antigens which used to be protective against a given pathogen. Therefore, when the population is exposed anew to the pathogen after many generations, a disease which is quite benign in some groups might become mortal in the one which was isolated. The converse also holds.

There are several points which can be discussed profitably in this model. The first point is the following: In many populations there is a very large number of alleles. Are they all useful for resistance, or if they are not, why do they persist in the population? One would expect that only those alleles which are more useful are maintained in the population. If the new mutant is more advantageous than the old one, in the course of generations it would supplant the old one.

Another point seems important. The mutability of the pathogen per generation is the same as the mutability in man, but the generation time confers an enormous adaptive advantage to the pathogen, which can experiment with countless mutations in the time in which man can experiment with one. It would, therefore, seem that the adaptation of man to new mutations in the pathogen, which increase virulence and decrease resistance in man, is very slow. It might be cautiously advanced that it might be the higher mutability of the pathogen in absolute time which renders advantageous the persistence of most of the haplotypes in the population. A given allele might become neutral, but it might resume advantage in due course if kept in the population.

A point which seems to us of importance in advocating advantage for all new HLA mutants is the fact that the HLA loci are duplications of a proto HLA locus. The first 16 amino acids of the products of the A and B loci are equal in sequence, and this speaks strongly in favor of duplication. If there are more loci at the HLA region, and most originate from duplications which did and can evolve independently of each other to a large extent, then the variability at the haplotype level is practically infinite and each individual will have a different HLA haplotype and phenotype, which will confer on the individual a basic resistance toward those pathogens which need membrane interactions to infect a host. Only a minority of humans will happen, by chance, to offer good adaptation to the pathogen.

It might be advanced, at this stage, that the theory that a new mutant is advantageous may be further refined, in the sense that the new allele which appears by mutation is not only advantageous of other aleles in haplotype combinations particularly, if the membrane surface specificity is the product of the activity of more than one locus. If, when one allele appears, the fitness of the others is increased, then it would be very simple to justify the extreme variability of the HLA polymorphism. It would be maintained by mechanisms of disease resistance which basically could represent a defense against the rapid adaptability of pathogens.

## REFERENCES

1. Barrai, I.: Spontaneous mutation rates in man. Biol Zentralblatt (in press), 1977.

2. Belvedere, M., Bernoco, D., Castagneto, M., Contu, L., Curtoni, E.S., Gandini, E., Mayr, W.R., Mattiuz, P.L., Reali, V., Savi, G., and Sirchia, G.: Colloborative participation in the sixth International Histocompatibility Workshop. I. The SD data. Histocompatibility Testing, Munksgaard, Copenhagen, 1975, pp. 115-121.

3. Bodmer, W.F.: Evolutionary significance of the HL-A system. Nature 237: 139-145, 1972.

4. Bodmer, W.F.: Evolution of HLA and other major histocompatibility systems. Genetics 79: 293-304, 1975.

5. Bodmer, W.F., Cann, H., and Piazza, A.: Differential genetic variability among polymorphisms as an indicator of natural selection. Histocompatibility Testing, 1972, Munksgaard, Copenhagen, 1973, pp. 753-757.

6. Bodmer, J.G., Bodmer, W.F., Pickbourne, P., Degos, L., Dausset, J., and Dick, H.M.: Combined analysis of three studies of patients with Burkitt's lymphoma. Tissue Antigens 5: 63-68, 1975.

7. Bodmer, W.F., and Thomson, G.: Population genetics and evolution of the HLA system. HLA and Disease, Munksgaard, Copenhagen, 1976.

8. Hill, H.F.H., Hill, A.G.S., and Bodmer, J.G.: Clinical diagnosis of ankylosing spondylitis in women in relation to presence of HLA-B27. Ann Rheum Dis 35: 267-270, 1976.

9. Kimura, M.: Random genetic drift in multi-alleleic locus.
<u>Evolution 9: 419-435, 1955</u>.

10. Mattiuz, P.L., Ihde, D., Piazza, A., Cappellini, R., and Bodmer, W.F.: New approaches to the population genetics and segregation analysis of the HL-A system.
<u>Histocompatibility Testing 1970</u>, Munksgaard, Copenhagen, 1970, pp. 193-205.

11. McDevitt, H.O., and Bodmer, W.F.: HL-A, Immune response genes, and disease.
<u>Lancet, 1269-1275, 1974</u>.

12. Svejgaard, A., Jersild, C., Nielsen, L. Staub, and Bodmer, W.F.: HL-A antigens and diseases. Statistical and genetical considerations.
<u>Tissue Antigens 4: 95-101, 1974</u>.

# *Psychophysiological and Neurochemical Analysis of Mnesic Processes*

Bernard Cardo

M. Claude Destrade

Alix Ebel

M. Robert Jaffard

The results summarized in the present report are located at the crossroads of several current preoccupations of psychophysiology.

First, data furnished by human pathology, and to a lesser degree those resulting from animal experimentation (19), have implicated hippocampal participation in mnesic processes.

Secondly, the role of central cholinergic systems in the regulation of behavior has been the object of a considerable amount of work and of various hypotheses (5,17,30). Considering the importance of hippocampal cholinergic systems, the second perspective often overlaps the first.

Finally, the use of different inbred strains of mice for behavioral as well as for neurochemical and pharmacological studies has led to the establishment of certain correlations between these two series of variables (41).

In our own experiments we first compared the performance of three inbred strains of mice in learning situations using the same reinforcement. We then established a correlation between behavioral results and the activity of hippocampal cholinergic mechanisms. Finally, in a more psychophysiological perspective, we tested the effects of limited hippocampal interventions both on the behavioral parameters studied and on hippocampal cholinergic activity.

---

This work is supported by the C.N.R.S. (ERA N$^{O}$ 416 and ATP Grant "Petit Mammifere")

## INTERSTRAIN COMPARISONS IN DIFFERENT LEARNING TESTS WITH POSITIVE REINFORCEMENT

In all the experiments summarized here, male mice belonging to lines BALB/c Orl, C57BL/6 Orl, and C57BR Orl were used. The animals were kept in collective cages in a controlled environment animal house (21-22°C) until the age of 8 weeks, at which time they were placed in individual cages with food and water *ad libitum*. The experiments were begun when the animals were 12 weeks old. Since all the experiments performed involved food reinforcement, the mice were progressively food-deprived before the tests so that at the time of the experiment their weight loss represented 16-19% of the *ad lib*. body weight (20,21).

### Study of the Rate of Acquisition

This first experiment was aimed at studying the rate of acquisition of a bar-press response on continuous reinforcement (CFR) during a single uninterrupted session in a Skinner box. The cage contained a lever and a food cup separated by a small partition which forced the animal to make a slight detour either to push the lever or to eat (11). Each lever-press freed a 6 mg food pellet in the cup. Twenty-seven mice of each line were used. No animal underwent a pretraining session and each animal was left in the apparatus for 30 minutes. The number of responses, defined as pushing the lever and then taking food, was recorded at 5 minute intervals.

The results (Figure 1) show that C57BR animals are quicker than C57BL/6 and BALB/c to associate the lever-press with food reward. The slowest rate of acquisition is observed for BALB/c, while that of C57BL/6 mice is intermediate. Calculation of the number of responses required before reaching the criterion of 10 responses per five minutes yields $25.8 \pm 1.1$ for BALB/c, $19.9 \pm 1.2$ for C57BL/6 and $11.\overline{6} \pm 1.2$ for C57BR ($p < 0.001$ for all statistical comparisons).

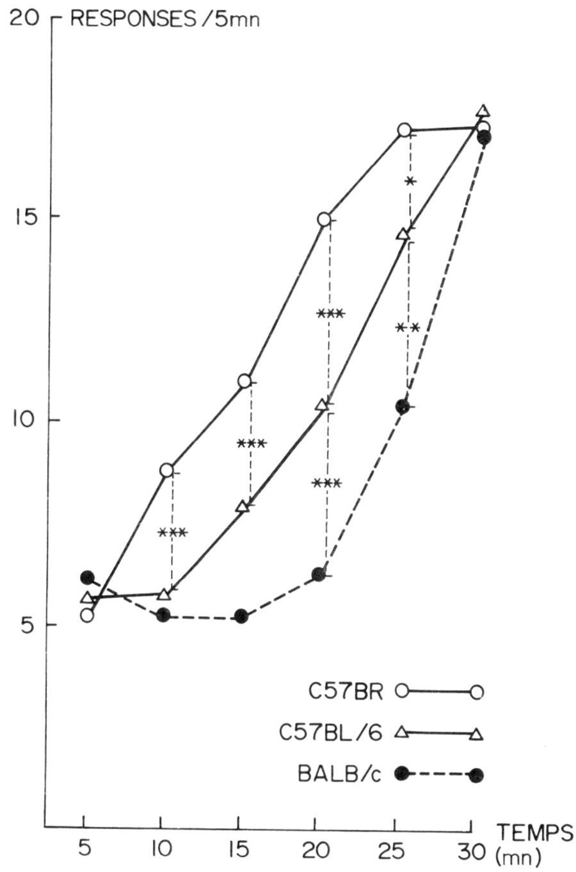


Figure 1. Rate of learning to push on a lever under conditions of continuous alimentary reinforcement in BALB/c, C57BL/6, and C57BR mice. Average performances (ordinate) recorded by successive 5-minute intervals (abscissa) during a 30-minute session. Statistical significance: * : $<$ 0.05; * * : $<$ 0.01: * * * : $<$ 0.001.

## Influence of the Retention Interval

We wished, in this second experiment, to test the effect of a retention interval occurring between an initial short session and a second session. Twenty mice of each line were prepared as described above and initially underwent a 30-minute session limited to 16 responses. Subsequently, 10 animals of each line underwent a 30-minute session which started 2.5 minutes after the end of the first session; the remaining 10 mice of each line underwent the same second session 24 hours later. The later animals were fed in such a way that their weight did not vary during the 24-hour interval.

The results are summarized in Figure 2. As in the preceding experiment, C57BR mice are superior to BALB/c and C57BL/6 during the initial session. In the second session, following a short retention interval, the order of performance of the three strains remains unchanged. After a 24-hour retention interval, however, the order is inverted. For BALB/c mice, the performance during the second session was significantly improved after an interval of 24 hours in comparison with their performance after an interval of 2.5 minutes ($F(1,18) = 12.23$; $p < 0.005$). The introduction of a 24-hour interval between tests reduced the performance of C57BL/6 animals ($F(1,18) = 6.15$; $p < 0.025$). A similar but nonsignificant phenomenon was observed with C57BR animals.

These two experiments demonstrate an inverse relationship between rate of acquisition and influence of the retention interval. The strains (i.e., C57) which exhibited a faster rate of acquisition when testing was uninterrupted, or interrupted only briefly, were disturbed when the interval between sessions was increased to 24 hours. Conversely, the animals (BALB/c) which exhibited the slowest acquisition under continuous testing showed a great improvement of performance after a 24-hour retention interval.

Nevertheless, the Skinner box test is relatively simple; the animal receives multiple information. Moreover, the permanence of the reinforcement permits no analysis of the behavior inhibition capacities of the mice. For this reason we compared the performance of the three

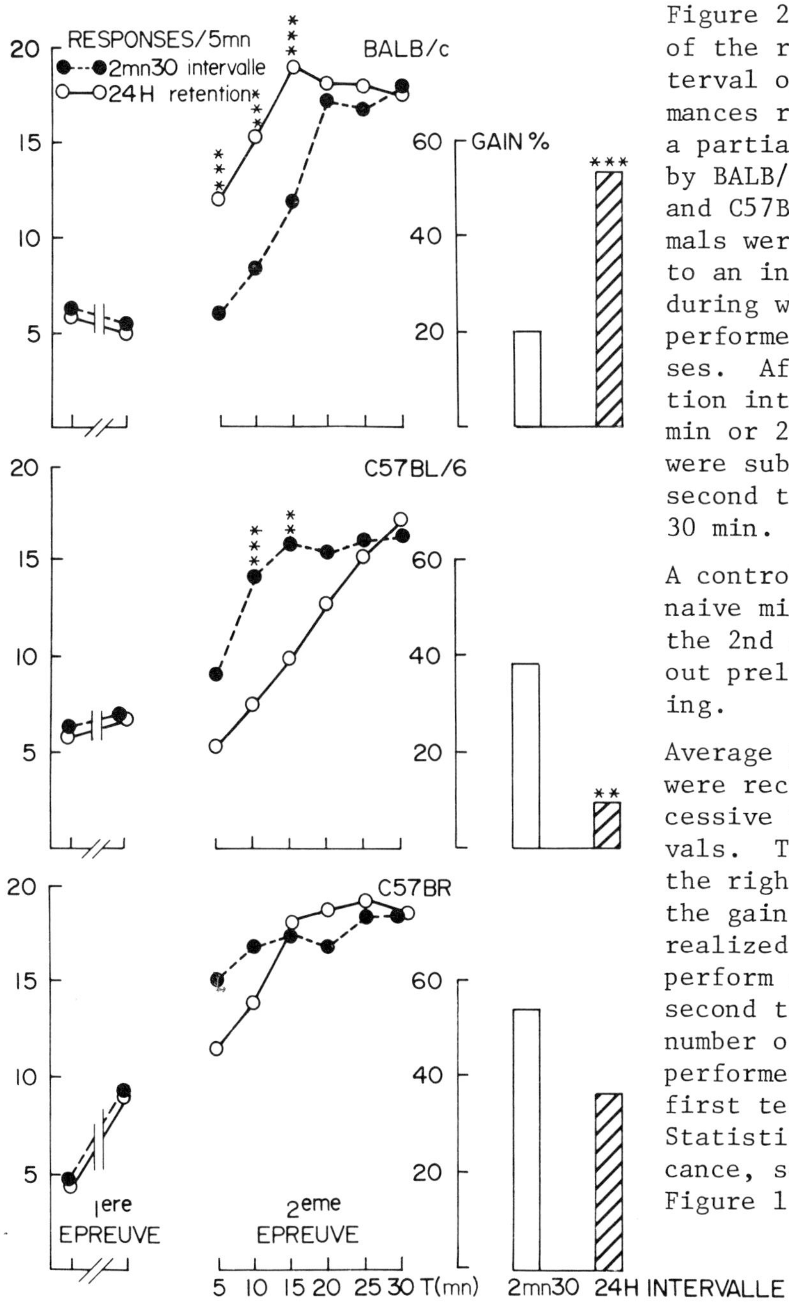


Figure 2. The effect of the retention interval on performances recorded after a partial acquisition by BALB/c, C57BL/6, and C57BR. The animals were subjected to an initial test during which they performed 16 responses. After a retention interval of 2.5 min or 24 hrs they were subjected to a second test lasting 30 min.

A control group of naive mice underwent the 2nd session without preliminary training.

Average performances were recorded by successive 5-min intervals. The columns on the right indicate the gain in time (%) realized in order to perform during the second test the same number of responses performed during the first test (16). Statistical significance, see legend to Figure 1.

lines in two other tests in order to be able to generalize our observations.

## The Extinction Test

Twenty mice from each of the three strains initially underwent a continuous reinforcement acquisition for 30 minutes, followed by two 15-minute retention sessions 24 and 48 hours later. The animals were then fed *ad libitum* for four days and were once again deprived according to the normal procedure. The animals then underwent four 10-minute extinction sessions separated by a 24-hour interval. This protocol allowed us to compare performances after short interval tests (the first five minutes compared to the last five minutes of the same session) to performances separated by the 24-hour retention interval (last five minutes of one session compared to the first five minutes of the following session).

The results (Figure 3) indicate that animals of the BALB/c line exhibit their greatest improvement of performance when testing is interrupted by a 24-hour retention interval, while animals of the C57 lines improve most dramatically in the course of a single session.

## The Acquisition of a Discriminative Operant Learning

Once again 20 mice from each line were used. They were prepared as before and subjected to three sessions of continuously reinforced learning as in the previous experiment; then they were subjected to discriminative learning eight days later. The apparatus consisted of the usual Skinner box, but release of food occurred only during certain periods marked by a light stimulus (25 lux) and a 70 dbA buzzer, both of which lasted the duration of the reinforcement period (SD). These signals were interrupted during all non-reinforcement periods (S$\Delta$). Each learning sequence, which lasted for 10 minutes, consisted of six reinforced S$\Delta$ periods of varying length (average

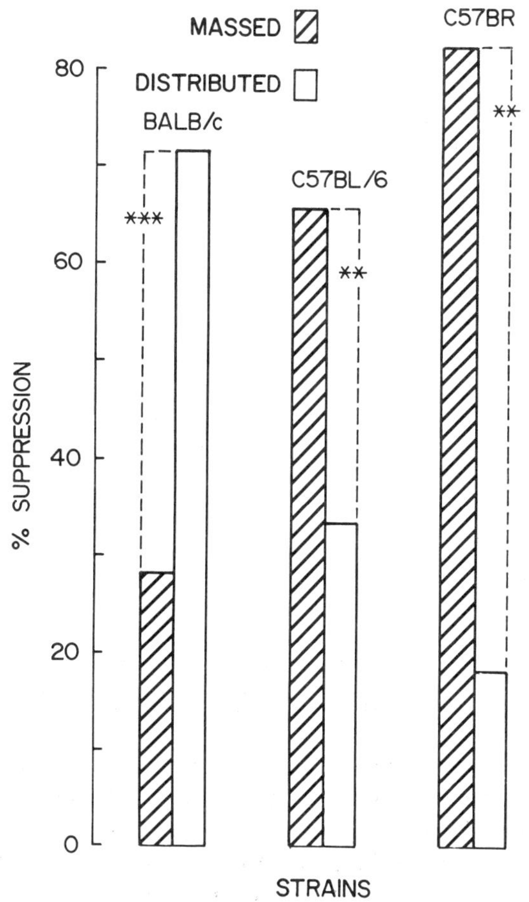


Figure 3. Extinction of an alimentary learning in a Skinner Box. The percentage of suppression (ordinate) for each strain during massed or distributed trials (abscissa). Statistical significance, see legend to Figure 1.

50 seconds) and six nonreinforced S$\Delta$ periods of the same average length. A session lasted 20 minutes and consisted of two 10-minute sequences. The animals underwent four sessions, separated by intervals of 24 hours.

As in the preceding experiment, we can compare the improvement registered between consecutive 10-minute tests with the improvement between tests separated by a 24-hour interval (21). The results indicate that there is no difference between the three strains at the level of SD response, regardless of the temporal spacing of the tests. Figure 4 shows the suppression of the responses during the nonreinforcement periods (SΔ) and indicates that this suppression in BALB/c mice predominates during the 24-hour intervals, in opposition to the C57 animals, which suppress their responses during the massed sections.

In conclusion, we have observed that the BALB/c mice are disadvantaged during a continuous acquisition, whereas the C57 mice are favored. Conversely, when a 24-hour retention interval is interposed, the performance of the BALB/c mice is considerably improved compared with the disturbed performance of the C57 animals.

## EFFECTS OF HIPPOCAMPAL STIMULATION OF THE PERFORMANCES OF THE THREE LINES

Some of our previous experiments have shown that a mild stimulation of the dorsal hippocampus performed immediately after the end of a partial learning led to a significant improvement of those performances 24 hours later in BALB/c (7,10-12, 39). In other words, stimulation of this strain further increased the natural gain related to the retention interval. It was thus important to know the effects of this same stimulation on the two other strains, which do not exhibit a natural gain. Two hypotheses can be formulated: hippocampal stimulation would induce an artificial gain and thus the mechanisms which exist in BALB/c probably also exist in the C57 strains, but are not expressed; or the stimulation would induce no effect, which would demonstrate the absence of the factor(s) implicated in the process (9).

Twenty mice from each strain were heavily anesthetized with sodium pentobarbital (100 mg/kg

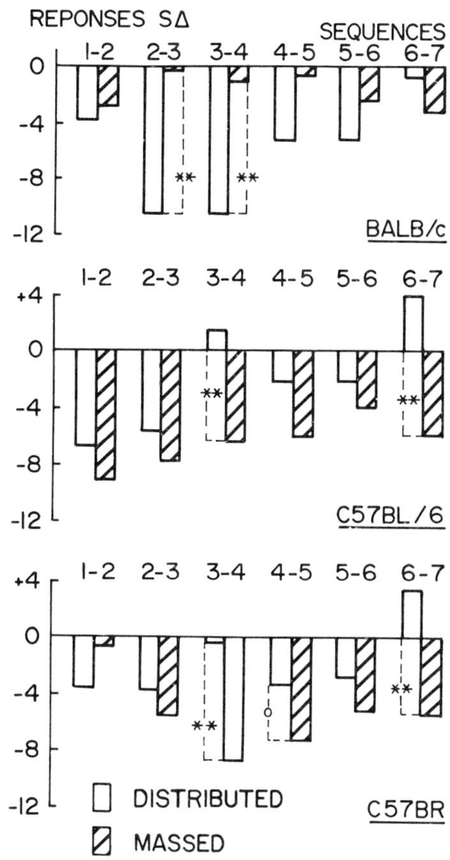


Figure 4. Acquisition of a discriminative operant learning. For each strain, two groups of mice were formed: one began the learning by a 10-minute sequence followed by three 20-minute sessions at 24-hour intervals. The other group began a 20-minute session, followed by two 20-minute sessions at 24-hour intervals and finally by a 10-minute session. This protocol enables the expression of the percentage of responses suppressed (ordinate) between two sequences, either massed or distributed (abscissa). Statistical significance, see legend to Figure 1.

for BALB/c, 70 mg/kg for C57BL/6 and 45 mg/kg for C57BR). A bipolar electrode was then inserted in the CA1 field of each dorsal hippocampus. Each electrode was made of two twisted irridiated platinum wires (diameter 0.09 mm), sheathed and sectioned at the tip.

Following postoperative recovery the threshold of hippocampal post-charges was determined for each animal (100 Hz sinusoidal stimulation for four seconds) according to a previously described technique (11). The animals of each strain were then deprived of food according to the normal procedure and subjected to a partial acquisition session (16 responses) in the Skinner box. Thirty seconds after the end of this session, 10 mice from each strain received a bilateral stimulation whose intensity was half the threshold value and which lasted 80 seconds (group ST). The remaining 10 mice of each group were not stimulated (group NST). Both groups of animals were then subjected to a second session lasting 30 minutes, 24 hours later. No significant difference was observed in the threshold of postdischarge triggering. Thus, the half-threshold values utilized during the experiments are quite similar: $m = 8.6 \pm 0.4$ $\mu$A for BALB/c, $8.5 \pm 0.3$ $\mu$A for C57BL/6 and $7.6 +$ $\mu$A for C57BR.

The results are summarized in Figure 5. The ST mice of the BALB/c strain exhibit performances which are superior to those of the NST mice, for the ensemble of the second session ($F(1,18) = 13.84$; $p < 0.005$). Facilitation is observed beginning with the first five minutes ($t(18) = 6.63$; $p < 0.001$).

A slight facilitation of stimulate C57BL/6 animals was observed ($F(1,18) = 4.58$; $p < 0.05$) but no effect was observed at the beginning of the test.

Finally, there was no effect of stimulation on C57BR mice.

Thus, post-trial hippocampal stimulation effects are strain-dependent and only serve to optimalize the previously observed differences on a 24-hour delayed retention test.

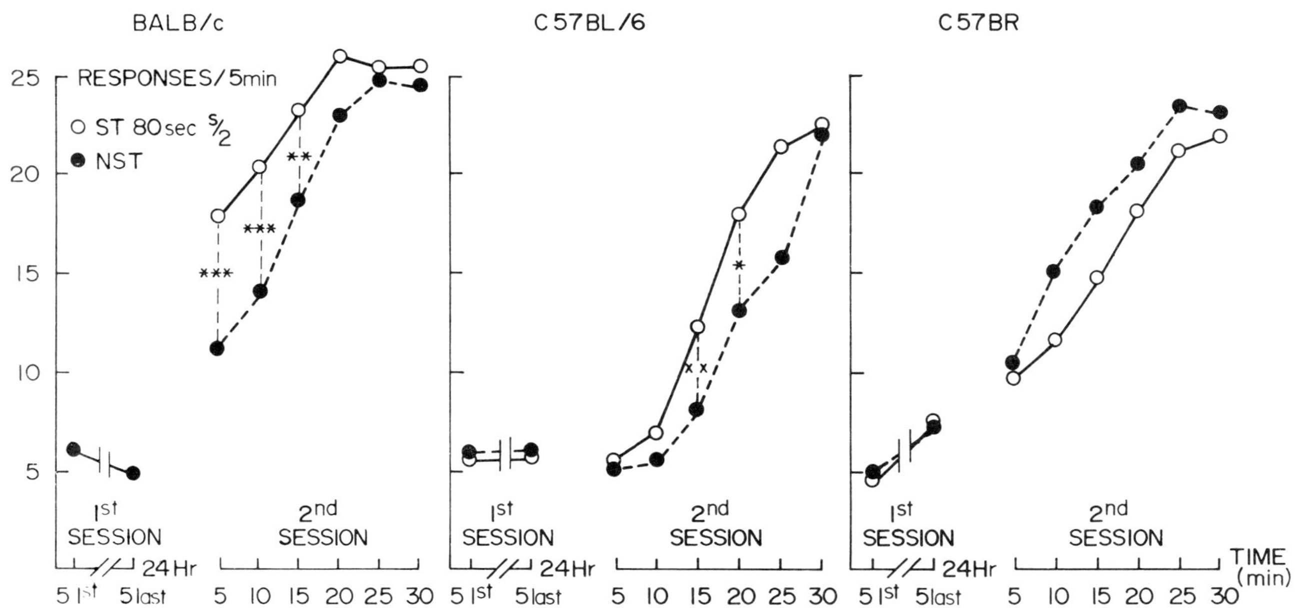


Figure 5. Modifications of a bilateral electrical stimulation of the dorsal hippocampus, applied for 80 seconds, 30 seconds after the first test, on performances recorded 24 hours later for BALB/c, C57BL/6, and C57BR mice. The average performances were recorded during successive five-minute intervals. Statistical significance, see legend to Figure 1.

## POSSIBLE ROLE OF THE HIPPOCAMPAL CHOLINERGIC SYSTEMS

Considering the region stimulated, on the one hand, and the supposed importance of the hippocampus and its cholinergic systems in behavioral regulation, on the other hand, we decided to investigate the eventual interstrain differences at the level of these systems (23). Several experiments were performed to this effect.

1. In an initial experiment, we attempted to differentiate the three strains by studying the activities of the enzymes involved in the synthesis and degradation of acetylcholine in the dorsal hippocampus.

Twelve mice of each of the three strains were subjected to isolation and progressive food deprivation according to the same protocol as used for the behavioral studies. After sacrifice of each animal by decapitatio, the two dorsal hippocampi were immediately removed and frozen in dry ice. The samples were homogenized at 0.2$^{o}$C in 0.5% Triton X-100 (100 $\mu$g tissue/1 $\mu$l); they were then diluted with 0.05% serum albumin. The activities of choline acetyltransferase (ChAc, E.C. 2.3.1.6) and acetylcholinesterase (AChE, E.C. 3.1.1.7) were determined using the isotopic microtechniques of McCaman *et al*., (32,33).

Table 1 shows that ChAc is more active in the dorsal hippocampus of BALB/c than in those of the two other strains, C57BL/6 ($t(22)=5.61$; $p < 0.001$) and C57BR ($t(22)=6.55$; $p < 0.001$). No important differences in AChE activites were observed (16, 24).

2. Since the preceding results appeared to us to be consistent with the behavioral data, we decided to test the effect of a half-threshold hippocampal stimulation on ChAc activity. Mice from the three strains (N=108 for BALB/c, N=48 for C57BL/6, N=24 for C57BR) were prepared under conditions strictly identical to those used in the experiment previously described (isolation, implantation, threshold determination, food deprivation). Mice assigned to the experimental groups

Table I

Comparison of the activities of the enzymes of synthesis (choline acetyltransferase ChAc) and degradation (acetylcholinesterase, AChE) of acetylcholine in the dorsal hippocampus of the strains BALB/c, C57BL/6 and C57Br. (1) $\mu$moles.g protein$^{-1}$.h$^{-1}$ (2) $\mu$moles.g protein$^{-1}$.h$^{-1}$ (3) mean ± S.E.M.

| STRAINS | | ChAc (1) | ACHE (2) |
|---|---|---|---|
| BALB/c | (N=12) | 15.32 ± 0.35 (3) | 4.15 ± 0.09 |
| C57BLc/c | (N=12) | 11.11 ± 0.51 | 3.88 ± 0.05 |
| C57BR | (N=12) | 10.22 ± 0.46 | 3.73 ± 0.07 |

were then subjected to electrical stimulation of the hippocampus for 80 seconds at the half-threshold level (ST groups); nonstimulated mice served as implantation controls (NST groups). Animals from the two groups were then sacrificed at varying intervals after stimulation (BALB/c: 5 and 30 min., 1,3, 6 and 24 hours and 6 days; C57BL/6: 3 and 24 hours and 6 days; C57BR: 3 hours and 6 days).

Stimulation of BALB/c mice induces important ChAc modifications, especially an increase after three hours (+ 47%; $p < 0.001$), followed by a decrease between six and 24 hours to a level inferior to the initial level ($p < 0.001$ in both cases). Activity is once again normal six days after stimulation. A slight increase is recorded in C57BL/6 after three hours (+ 21%; $p < 0.05$). No modifications are observed with C57BR. AChE activity is not affected in any of the strains.

3. If we compare previous data with the behavioral results summarized in § 1, the following correlates can be observed: In BALB/c mice,

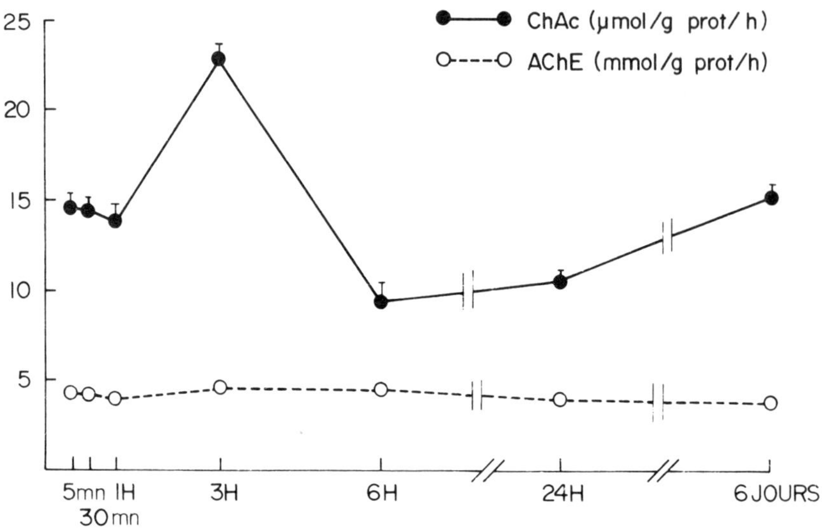


Figure 6. Modifications of acetylcholinesterase (AChE) and choline acetyltransferase (ChAc) activities induced at variable delays by electrical stimulation of the hippocampus of BALB/c.

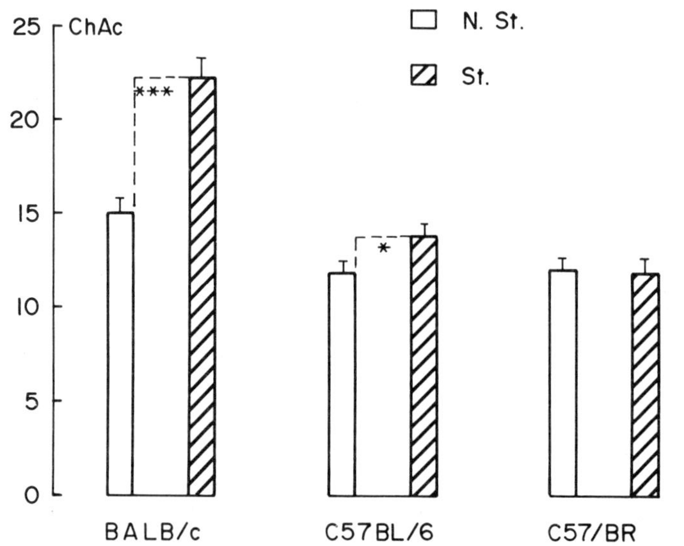


Figure 7. Modification of ChAc activity (as µ moles.g protein$^{-1}$.h$^{-1}$) by electrical stimulation of the hippocampus of BALB/c, C57BL/6, and C57BR when the treatment is applied three hours before removing samples. See legend to Figure 1 for statistical significance.

favored by the retention interval and disturbed by continuous acquisition, ChAc activity, and thus the activity of the cholinergic system, is important. On the contrary, in the two C57 lines which are favored by continuous acquisition and disturbed by the retention interval, ChAc activity is low. In addition, hippocampal stimulation, which favors the effect of the retention interval in the BALB/c strain, modifies ChAc activity only in this strain.

The hypothesis of a positive link between cholinergic activity and the effect of the retention interval, and of a negative link between this activity and massed acquisition can be formulated. In order to test this hypothesis, we used previous results which showed that 24 hours following the stimulation of BALB/c mice, ChAc activity was quite comparable to the natural activity of the C57 strains.

Forty-eight BALB/c mice were prepared under normal conditions (isolation, implantation, threshold determination, and food deprivation). In one group of 24 mice, seven received hippocampal stimulation (80 seconds at half-threshold value) 24 hours before an acquisition test (group ST 24), seven others received this same stimulation five minutes before the test (group ST 5 min) and the remaining 10 animals were not stimulated (NST). The three groups were then subjected to a continuous acquisition session lasting 30 minutes.

Table II shows the influence of the treatments of ChAc activity and also indicates the performances of the three groups, expressed as the number of responses necessary in order to attain the level of 10 responses per five minutes. It can be seen that the stimulation applied 24 hours before acquisition increases its rate ($p < 0.001$). The same stimulation applied five minutes before acquisition has no effect. The depression of ChAc activity which is artificially induced in BALB/c thus increases the rate of acquisition, which is normally disadvantaged in this strain.

In the second group of 24 mice, three groups were also formed (N=7 for ST 24, N=7 for ST five

Table II

Modifications of the rate of acquisition of alimentary conditioning in BALB/c by electrical stimulation of the dorsal hippocampus applied 24 hours (ST 24h) or five minues (ST five min) before the test. The group of nonstimulated animals (NST) serves as a control. The rate of acquisition is evaluated by the number of responses necessary in order to reach the level of 10 responses per five minutes. ChAc activity, expressed as moles.g protein$^{-1}$.h$^{-1}$, is recalled on the right hand portion of the table. All values are expressed as means ± S.E.M. (a) vs (b) and (c) : $p < 0.001$ (d) vs (e) and (f) : $p < 0.001$

| GROUPS | RESPONSES TO CRITERION | ChAc |
|---|---|---|
| BALB/c ST-24h | 16.9 ± 0.6 (a) | 10.65 ± 0.61 (d) |
| BALB/c ST-5mn | 26.5 ± 1.2 (b) | 14.51 ± 0.70 (e) |
| BALB/c NST | 26.4 ± 1.0 (c) | 15.06 ± 0.64 (f) |

minutes and N=10 for NST). They underwent an initial partial acquisition session, followed by a second session 24 hours later (cf. § 1,2).

It can be seen in Table III that group ST 24 exhibits a significant deficit at the beginning of the second session (comparison with ST five minutes and NST, $p < 0.001$ in both cases). In other words, the artificially induced ChAc decrease suppresses the effect of the retention interval.

## DISCUSSION

Our results can be discussed under three different headings.

1. The factor which differentiates the three strains in hunger-motivated learning situations is the temporal spacing of testing. More exactly, we

TABLE III

Modifications of retention capacities in BALB/c by electrical stimulation of dorsal hippocampus performed 24 hrs. or five mins before partial learning test. Performances (mean $\pm$ S.E.M.) recorded during last five mins of partial acquisition and during first five mins of retention test (24 hrs. later). (a) vs (b) and (c) : nonsignificant (d) vs (e) and (f) : $p < 0.001$.

| GROUPS | 1st SESSION (last 5 mn) | 2nd SESSION (first 5 mn) |
|---|---|---|
| BALB/c ST-24h | 5.4 ± 0.3 (a) | 5.4 ± 0.6 (d) |
| BALB/c ST-5mn | 5.1 ± 0.3 (b) | 11.0 ± 0.6 (e) |
| BALB/c NST | 5.6 ± 0.3 (c) | 11.0 ± 0.8 (f) |

forced the animals to process information either immediately, i.e., during the test, or external to the experimental situation. Whether it consists of a continuous reinforcement test, an extinction test optimalizing capacities for behavioral inhibition, or a discrimination test in which permissive periods are controlled by external stimuli, under the above mentioned conditions a long retention interval improves the performances of BALB/c mice while perturbing those of C57 animals. In the absence of this interval, however, the C57 strains perform better than BALB/c. Furthermore, the influence of the 24-hour retention interval in BALB/c is manifested as an important gain beginning with the start of the second session. This gain, which can be compared with the phenomenon of reminiscence, well known in man and to a lesser degree in animals (21), shows that this strain can process some stored information. It is this delayed information processing which is apparently lacking in the C57 strains. The absence of a hippocampal stimulation effect in these mice confirms the importance of this deficit.

2. In terms of an explanation of the observed differences, we can note a) that the hippocampal cholinergic system of BALB/c is more active than those of the C57 strains, b) that hippocampal stimulation has important repercussions on the cholinergic system only in BALB/c mice, and c) that if a decrease of the activity of this system is induced in BALB/c, modalities of acquisition and retention are induced which are close to those observed for C57BL/6.

Taken together, these three arguments seem to indicate that the hippocampal cholinergic system modulates learning mnesic operations, leading in particular to delayed processing of data received.

This is not a novel idea. A considerable number of authors, using either a pharmacological (5,17) or a neurochemical (28-31) approach, have shown the importance of central cholinergic systems in learning mechanisms. In particular, Matthies (31) demonstrated a considerable increase in the level of free acetylcholine in the rat hippocampus immediately after a session of discriminative avoidance conditioning. It is probable that under our conditions, the increase of ChAc activity observed three hours after stimulation of BALB/c is only a delayed secondary effect of previous operations. Indeed, we could show that hippocampal stimulation facilitated performance 24 hours later only if it was applied less than 10 minutes after the end of the partial learning (11).

Our current working hypothesis is thus the following: the end of partial learning in BALB/c would lead to an important activation of the hippocampal cholinergic system with an increased utilization of acetylcholine. Hippocampal stimulation would artificially increase this liberation which, provided it occurs within the several minutes following the session, would lead to an improvement of performance 24 hours later. In both cases, this liberation would subsequently lead to an increase in ChAc activity. It is thus probably the temporal association between the end of the registration of information and the liberation of acetylcholine which represents the fundamental phenomenon.

Because of the weak importance of the hippocampal cholinergic system in the two C57 strains, this liberation of acetylcholine would either not occur at all or would be too weak or delayed.

This cholinergic system is well known by neurochemists; it includes the septo-hippocampal afferent pathways, considered to be mainly cholinergic (27). It is important to further explore the interstrain differences observed: the differences in ChAc activities do not tell us which real differences exist among the three strains at the septo-hippocampal level. We, nevertheless, have certain results at our disposition concerning interstrain differences at the neuro-anatomical level. Wimer _et al._ (43) compared the hippocampal volumes of nine inbred strains of mice and found that the hippocampus of BALB/c had the largest volume. In addition, the intrahippocampal organization of this strain seems to be different from that of the other strains analyzed (3). This last finding might explain the neurochemical differences presently observed by us.

3. Finally, we address ourselves to the functional significance of the observed differences. The disagreement in this respect is important.

For some authors (13), central cholinergic systems have a predominant role in the consolidation of stored information. Certain results, however, do not argue in favor of this hypothesis (35,38). Furthermore, our results show that the hippocampal cholinergic system intervenes inversely both in immediate and delayed information processing: the BALB/c mouse, whose cholinergic system was artificially depressed, improves its performances during immediate processing (continuous acquisition), but on the contrary exhibits no improvement after a 24-hour interval. This induced double action of the cholinergic system leads one to think that it intervenes as a secondary nonmnesic variable.

For others, the hippocampal cholinergic systems would have behavioral inhibitory functions. There are two origins of this hypothesis: the hippocampus is often considered as an inhibitory structure (2,14,19,26); in addition, based on pharmacological arguments, numerous authors admit the inhibitory function of central cholinergic systems (5,6,15). Unfortunately, the hypothesis according to which the hippocampus would have inhibitory functions has been considerably criticized (4,34), as has been the hypothesis attributing these functions to cholinergic systems (1,37,42). Furthermore, our results show that C57 mice can inhibit a behavior pattern previously acquired during massed learning. The deficit is not at the level of the inhibition, but rather at the level of the distinction between immediate and delayed inhibition.

Numerous behavioral and neurophysiological data confirm that hippocampal control is much more complex than believed. The least unsatisfactory hypothesis, but also the most difficult to explore, supposes that the hippocampus functions as an interface whose role is to compare behavioral sequences or programs to motivational data or to data involved with vigilance. It is a system whose function is the estimation of the valence of a program or plan and not its informational value. The role of the hippocampus in the regulation of motivated behavior is well known (25). In addition, studies of unitary hippocampal responses have confirmed the sensitivity of these cells in the habituation process (40) and in approach behavior (36).

This general concept would attribute a function of comparison between behavioral sequences and reinforcement processes, at least positive ones, to the hippocampus; it is fairly consistent with our results.

The gain observed in BALB/c after a 24-hour interval can only be explained by a selection operation leading to the elimination of nonreinforced responses and to the retention of those which were reinforced.

In addition, we could show that stimulation of the lateral hypothalamus in BALB/c, in the most efficacious positive reinforcement region, led 24 hours later to the same improvements as hippocampal stimulation and to a parallel decrease in hippocampal ChAc activity (8). This result clearly shows that the hypothalamo-septo-hippocampal pathway is involved. Considering the importance of the lateral hypothalamus in homeostatic regulations as well as in the process of positive reinforcement, we can admit that the natural phenomena observed in BALB/c, as well as the phenomena induced by stimulation, result from the operation of hypothalamic pathways directly or indirectly destined to the hippocampus.

In other words, we believe that the inter-strain differences observed derive from different capacities for the transfer information originating in reinforcement systems.

It should be recalled here that hippocampal stimulation of BALB/c is effective only if it is applied less than 10 minutes after the end of the partial learning. This critical period must correspond to the transfer of a signal from the hypothalamus to the hippocampus. This transfer must lead to the release of large quantities of acetylcholine and it can be admitted as a working hypothesis that the triggering signal is nothing more than the end of the test.

Immediate processing by the C57 strains would not lead to the utilization of these hippocampal circuits since the confrontation between a motor sequence and reinforcement, or its absence, is immediate and demands no long-term memorization. With delayed processing, however, information coming from reinforcement systems could not be transferred toward the hippocampal circuits because of the low degree of projections toward these structures, particularly the cholinergic septo-hippocampal projections.

The final question to be raised is the possibility of an antagonism or a certain incompatability between delayed and immediate processing. Considering the very tight bonds between the hypothalamus and the hippocampus, it can be supposed that the development of an efficient hippocampal control system would limit hypothalamic capacities. It is not very probable that these limits are expressed only by inhibitions. Our results show, on the contrary, that whatever the behavioral response required (activation or inhibition), one of the main important factors which differentiates BALB/c and C57 performances is the temporal spacing of the retention tests.

## REFERENCES

1. Abeelen, J.H.F., van Ellenbrock, G.A., and Wigman, H.F.A.J.: Exploratory behavior in two selected-bred lines of mice after intrahippocampal injection of methylscopolamine. Psychopharmacologia 41: 111-112, 1975.

2. Altman, J., Brunner, R.L., and Bayer, A.: The hippocampus and behavioral maturation. Behav Biol 8: 557-596, 1973.

3. Barber, R.P., Vaughn, J.E., Wimer, R.E., and Wimer, C.C.: Genetically-associated variations in the distribution of dentate granule cell synapses upon pyramidal cell dendrites in mouse hippocampus. J Comp Neurol 156: 417-434, 1974.

4. Black, A.H.: Hippocampal electrical activity and behavior. In The Hippocampus, R.L. Isaacson and K.H. Pribaum (Eds.), Volume 2, Plenum Press, 1974, pp. 129-167.

5. Brimblecombe, R.W.: Drug Actions on Cholingeric Systems, P.B. Bradley (Ed.), Mc Millan Press, 1974, p. 227.

6. Carlton, P.: Cholinergic mechanisms in the control of behavior by the brain. Psychol Rev 70: 19-39, 1963.

7. Destrade, C., and Cardo, B.: Effects of post-trial hippocampal stimulation on time-dependent improvement of performance in mice. Brain Res 78: 447-454, 1974.

8. Destrade, C., and Cardo, B.: Amélioration de la réminiscence par stimulation post-essai de l'hypothalamus latéral chez la souris BALB/c. C R Acad Sci, Paris, 280D, 1401-1404,1975.

9. Destrade, C., and Jaffard, R., Deminiere, J.M., and Cardo, B.: Effets de la stimulation de l'hippocampe sur la réminiscence chez deux lignées de souris. <u>Physiol Behavior 16: 237-243, 1976</u>.

10. Destrade, C., and Jaffard, R.: Effets de la stimulation de l'hippocampe sur un apprentissage opérant discrimatif chez la souris BALB/c. <u>Neuropsychol, in press</u>.

11. Destrade, C., Soumireu-Mourat, B., and Cardo, B.: Effects of post-trial hippocampal stimulation on acquisition of operant behavior in the mouse. <u>Behav Biol 8: 713-724, 1973</u>.

12. Destrade, C., Soumireu-Mourat, B., and Cardo, B.: Influences de l'intensité de la stimulation hippocampique sur un apprentissage d' approche chez la Souris. <u>C R Acad Sci Paris 277: 225-228, 1973</u>.

13. Deutsch, J.H.: The cholinergic synapse and the site of memory. In <u>The Physiological Basis of Memory</u>, J.H. Deutsch (Ed.), Academic Press, 1973, p. 439, 59-76.

14. Douglas, R.J.: Pavlovian conditioning and the brain. In <u>Inhibition and Learning</u>, R.A. Boakes and M.S. Halliday (Eds.), Academic Press, London, 1972, pp. 529-533.

15. Douglas, R.J.: The development of hippocampal function: implications for theory and for therapy. In <u>The Hippocampus</u>, R.L. Isaacson and K.H. Pribram (Eds.), Volume 2, Plenum Press, 1975, pp. 327-361.

16. Ebel, A., Jaffard, R., Destrade, D., Ayad, G., Mandel, P. and Cardo, B.: Modification des mécanismes cholinergiques et de l'apprentissage chez trois lignées de souris consanguines par stimulation électrique de l'hippocampe dorsal.
C R Acad Sci Paris 283: 535-538, 1976.

17. Feudis, FV.de.: Cholinergic roles in motivated behaviours, emotion, learning and memory. In Central Cholinergic Systems and Behaviour, F.V. de Feudis (Ed.), Academic Press, London, New York, San Francisco, 1974, pp. 75-117.

18. Isaacson, R.L., and Kimble, D.P.: Lesions of the limbic system: their effects upon hypotheses and frustration..
Behav Biol 7: 767-793, 1972.

19. Isaacson, R.L.: Hippocampal destruction in man and other animals..
Neuropsychol 10: 47-64, 1972

20. Jaffard, R., Deminiere, J.M., and Cardo, B.: Etude comparée des capacités de rétention d'un apprentissage alimentaire chez deux souches de souris consanguines.
C R Acad Sci Paris 278: 2549-2551, 1974.

21. Jaffard, R., Destrade, C., Soumireu-Mourat, B., and Cardo, B.: Time-dependent improvement of performance on appetitive tasks in mice.
Behav Biol 11: 89-100, 1974.

22. Jaffard, R., Destrade, C., and Cardo, B.: Effets de la stimulation de l'hippocampe sur l'évitement passif chez deux lignées de souris.
Physiol Behav 16: 233-236, 1976.

23. Jaffard, R., Ebel, A., Destrade, C., Ayad, G., Mandel, P., and Cardo, B.: The role of hippocampal cholinergic mechanisms in the acquisition of bar-press response. Pharmacol Biochem Behav 5: 371-374, 1976.

24. Jaffard, R., Ebel, A., Destrade, C., Durkin, T., Mandel, P., and Cardo, B.: Effects of hippocampal electrical stimulation on long-term memory and on cholinergic mechanisms in three inbred strains of mice. Brain Res, in press.

25. Jarrard, L.E.: The hippocampus and motivation. Psychol Bull 79: 1-12, 1973.

26. Kimble, D.P.: Hippocampus and internal inhibition. Psychol Bull 70: 285-295, 1968.

27. Kuhar, M.J.: Cholinergic neurons: septal-hippocampal relationships. In *The Hippocampus*, R.L. Isaacson and K.H. Pribram (Eds.), Volume 1, Plenum Press, 1975, pp. 269-284.

28. Mandel, P., Ebel, A., Mack, G., and Kemf, E.: Neurochemical correlates of behavior in inbred strains of mice. In *The Genetics of Behavior*, J.H.F. van Abeelen (Ed.), North-Holland Publishing Company, 1974, p. 450, pp. 397-415.

29. Mandel, P., Ayad, G., Hermett, J.C., and Ebel, A.: Correlation between choline acetyltransferase activity and learning ability in different mice strains and their offspring. Brain Res 72: 65-70, 1974.

30. Matthies, H.: The biochemical basis of learning and memory. Life Sciences 15: 2017-2031, 1974.

31. Matthies, H., Ranca, C., and Liebmann, H.: Changes in the acetylcholine content of different brain regions of the rat during a learning experiment.
J Neurochem 23: 1109-1113, 1974.

32. McCaman, R.E., and Hunt, J.M.: Microdetermination of choline acetylase in nervous tissue.
J Neurochem 12: 253-259, 1965.

33. McCaman, R.W., Tomey, C.R., and McCaman, R.E.: Radiometric essay of acetylcholine sterase activity in submicrogram amounts of tissue.
Life Sciences 7: 233-244, 1968.

34. Nadel, L., O'Keefe, J., and Balck, A.: Slam on the brakes: a critique of Altman, Brunner and Bayer's response-inhibition model of of hippocampal function.
Behav Biol 14: 151-162, 1975.

35. Oliverio, A.: Contrasting effects of scopolamine on mice trained simultaneously with two different schedules of avoidance conditioning.
Psychopharmacologia 11: 39-51, 1967.

36. Ranck, J.B., Jr.: Behavioral correlates and firing repertoires of neurons in the dorsal hippocampal formation and septum of unrestrained rats. In The Hippocampus, R.L. Isaacson and K.H. Pribram (Eds.), Volume 2, Plenum Press, 1975, pp. 207-244.

37. Ross, J.F., and Grossman, S.P.: Intrahippocampal application of cholinergic agents and blockers.
J Com Physiol Psychol 86: 590-600, 1974.

38. Signorelli, A.: Influence of physotigmine upon consolidation of memory in mice.
J Comp Physiol Psychol 90: 658-664, 1976.

39. Soumireu-Mourat, B., Destrade, C., and Cardo, B.: Effects of seizure and subseizure post-trial hippocampal stimulation on appetitive operant behavior in mice. Behav Biol 15: 303-316, 1975.

40. Vinogradova, O.S.: Functional organization of the limbic system in the process of registration of information: facts and hypotheses. In *The Hippocampus*, R.L. Isaacson and K.H. Pribram (Eds.), Volume 2, Plenum Press, 1975, pp. 3-69.

41. Wahlsten, D. Genetic experiments with animal learning: a critical review. Behav Biol 7: 143-182, 1972.

42. Warburton, D.M.: The cholinergic control of internal inhibition. In *Inhibition and Learning*, R.A. Boakes and M.S. Halliday (Eds.), Academic Press, London, 1972, pp. 431-460.

43. Wimer, R.E., Wimer, C.C., and Roderick, T.M.: Genetic variability in forebrain structures between inbred strains of mice. Brain Res 16: 257-264, 1969.

# *Paramecium as a Model for Genetic Approaches to Neurobiology*

Roger Eckert

The functioning of the nervous system is based to a considerable extent on the electrical behavior of the surface membranes of nerve cells, an area of neurobiology that has been studied intensively for the past 35 years. Despite a wealth of data very little is known of the underlying molecular mechanisms, especially those which determine the passive permeabilities of membranes to ions. These poorly understood mechanisms are responsible for such phenomena as resting and action potentials, receptor and synaptic potentials, and most of the other nonmetabolic nerve and muscle functions such as the coupling of transmitter release and muscle contraction to changes in membrane potential. In all oı these the membrane modulates the passive movement of each species of ion by changing its conductance to that species. A number of separable factors enter into these conductance changes including the opening of hypothetical "gates," the subsequent closing of these gates, and voltage and time dependence of the opening and closing kinetics. Each gated channel shows a preference of "selectivity" for some ions over others. In electrically excitable membranes this selectivity is a function of membrane potential and

---

The work from my laboratory described herein was done with the invaluable collaboration of Y. Naitoh, C. Kung, and H. Machemer, and was supported by grants NSF BMS 74-19464 and USPHS NS 8364.

time. In receptor membranes it is modulated by stimulus energy such as light, heat, mechanical displacement, or stimulant molecules.

An understanding of membrane selectivity lies in the domain of molecular and submolecular mechanisms. Simplistic theories of selective permeability based on pore sizes and fixed charges have been upgraded by a universal theory for ion selectivity developed by Eisenman (10), and applied to biological materials by Diamond and Wright (2). Eisenman's theory for selectivity is based on physical-chemical interactions between sites on the membrane and the ions in solution. It states that the interaction between sites on the membrane and between an ion and a polar site is determined by the difference between the coulombic energy of attraction between the two and the hydration energy of the ion. Relative attractiveness of a site for two ionic species is therefore determined by the difference between these two net energies for the two species in question. Experimental studies are needed to determine relations between the actual molecular structure of cell membranes and their permeability channels and the chemical predictions that can be made from Eisenman's theory.

Among the most successful approaches to molecular problems in biology have been those that employ genetic manipulation of microorganisms. Microorganisms can be grown in cultures of genetically and phenotypically uniform populations uncontaminated by other cell types. They are readily modified genetically, and these modifications can be analyzed biochemically. Genetic modification (mutations) can be used to "dissect" sequential processes and structural organization. Analysis of genetically modified molecules should provide insight to the functions of the normal molecules. Genetic dissection has been successful in a number of systems such as sequential biosynthesis and phage assembly.

Genetic approaches to the nervous system are underway in several laboratories using fruitflies and roundworms. Behavioral tests have been devised to screen for neurally expressed mutations. Encouraging progress has already been made, but the analysis of these mutants at molecular levels is greatly complicated by the enormous structural complexity of even the simplest nervous systems. For this reason they are not ideally suited for analyses of membrane structure, except for special cases such as photoreceptor membranes of the compound eye. At the other end of the spectrum of behavioral genetics there are those, like Adler (1), who work with bacteria. Efforts over several years in Adler's laboratory to record membrane potentials from procaryote cells have failed to produce satisfactory results. It is unclear if bacterial cell membranes exhibit electric excitability. Bacterial membranes may be most useful for studies of steady state (resting) ion fluxes, ion pumping phenomena, and chemical transduction.

Between the hypercomplexity of the metazoan nervous system and the limited bioelectric interest and usefulness of bacteria lie the large, unicellular ciliated protozoa such as *Paramecium*. Some outstanding advantages presented by the ciliates for the genetic-biophysical approach to molecular mechanisms of bioelectric phenomena in membranes can be summarized as follows:

1. *Suitability for genetic modification and analysis*. There is a large genetic literature especially for *Paramecium aurelia*. This protozoan exhibits autogamy, which greatly facilitates the induction, selection, and genetic analysis of mutations.

2. *The locomotor behavior of ciliates is controlled by the electric behavior of the cell membrane*. For example, the wild type membrane is depolarized when the cation concentration is raised. This causes the cilia to beat in reverse, and the depolarization is detected as backward swimming of the specimen. This direct coupling of membrane behavior and locomotion is of great advantage because it allows detection of membrane mutants by overt behavior and permits screening procedures for selection of membrane mutants (15-17). This advantage is absent in the cultured neuroblastoma cells used by Nirenberg and Nelson.

3. The ciliates are unicellular, free of multiunit neural organization and complex synaptic integration. This characteristic eliminates the difficult search for modified units necessary in multicellular systems. It also avoids the statistical effects of organizational complexity on mutations which have a subtle diffuse effect on all the units of a complex system. It is well known from neuropharmacology that minute but widespread modifications of neuronal function can be greatly amplified by the integrative properties of the nervous system to produce behavioral effects far out of proportion to their cellular effects.

4. Ciliates such as *Paramecium* are large enough for routine intracellular recording, stimulating and clamping techniques. The cell is essentially isopotential, so that the location of the electrodes has little consequence and recording conditions are highly reproducible (6). There is a growing literature on their electrophysiology and the bioelectric control of locomotor behavior.

5. The cell membranes of ciliates exhibit the basic electrical properties found in nerve, muscle, and receptor cells of the metazoa. The membranes generate action potentials and receptor potentials. Ionic currents, especially the influx of calcium, regulate motor activities of the cell.

6. Ciliates can be handled with mass clonal culture technique for biochemical studies. Large numbers of genetically identical cells can be obtained without contamination by other eucaryote cell types.

The presence of an electrically excitable membrane in a microorganism suitable on the one hand for electrophysiological studies, and on the other hand for genetic and biochemical manipulation and analysis, presents a most promising opportunity for molecular and genetic approaches to the elucidation of membrane functions.

Electrophysiological analyses of the membranes of ciliated protozoa (*Paramecium* and *Euplotes*) have revealed electrical properties similar to those exhibited by receptor, nerve, and muscle cells of the metazoa. Mechanical stimulation of the surface membrane at the anterior end induces a depolarizing $Ca^{2+}$-mediated receptor potential, while similar stimulation of the rear end induces a $K^{+}$-mediated hyperpolarizing receptor potential. Polarization excites a "calcium system" in the cell membrane which is manifested in a graded regenerative "calcium response" with characteristics similar to the graded responses of crustacean muscle. $Ba^{2+}$ converts this response to all-or-none. The calcium response is followed after 2-8 msec by a reversal in the beating direction of the cilia. ATP-$Mg^{2+}$ reactivated cilia of detergent-extracted models of *Paramecia* shift into reverse when the ambient $Ca^{2+}$ concentration is raised to $10^{-6}$M or higher. Therefore, $Ca^{2+}$ which enters the cell during the calcium response is believed to activate the reversal mechanism. The locomotor behavior of the ciliate is based to a large extent on this sensory-motor sequence. The loss of the calcium response due to a single gene mutation in the mutant *Paramecium* "Pawn" renders this mutant unable to swim backwards in response to depolarizing stimuli. However, the reversal mechanisms remains functional, for removal of the membrane by a detergent renders the mutant sensitive to $Ca^{2+}$. Many of the data and concepts representative of the progress of the past several years are summarized in a series of review articles (3,4,7,9,23)

The electrophysiology of ciliated protozoa, previously an obscure curiosity far out from mainstream physiology, has now come to the attention of many orthodox neurobiologists because of our studies on 1) the role of the membrane in sensory transduction and 2) electro-mechanical coupling, 3) similarities of calcium response of the membrane to calcium responses of other membranes such as that of presynaptic nerve terminals, and 4) the demonstrated loss or alteration of this calcium system by a single gene mutation in *P. aurelia* (15,19). This new general interest *is* significant because it brings the genetic approach successfully into the general area of membrane physiology, and *Paramecium* becomes a model system for several basic neurobiological and membrane biophysical problems. The mechanisms regulating sensory and motor functions in a unicellular organism without synapses or nervous system are, of course, interesting in their own right as well.

Electric Properties of the Membrane (3,5,6,24). Potential recordings made simultaneously from opposite ends of the cell indicate that the cytoplasmic compartment of *P. caudatum* is nearly isopotential. Measured decrements of the spread of steady-state potentials were in essential agreement with calculated decrements for a short cable model of similar dimensions and electrical constants. Action potentials and passively conducted pulses spread at rates of over 100 μm per msec. In contrast, metachronal waves of ciliary beat progress over the cell with velocities below 1 μm per msec. Thus, electrical activity conducted by the plasma membrane cannot account for the metachronism of ciliary beat. The electrical properties of *Paramecium* are responsible, however, for coordinating the reorientation of cilia (either beating or paralyzed by $NiCl_2$) which occurs over the entire cell in response to depolarizing current passed across the plasma membrane. The demonstration that the cell interior is essentially isopotential is important for voltage clamp experiments.

Outward current delivered with a microelectrode produces a regenerative component superimposed on the passive upstroke. The regenerative depolarizations are graded in amplitude and rate of rise with the intensity of applied outward current pulses. The overshoot reaches an amplitude of +5 to +15 mV, and shows a maximum rate of rise of about 7V/sec. The overshoot is insensitive to $Na^+$, $Cl^-$, TTX and procaine, but increases slightly (up to 10 mV/ 10-fold rise in ion concentration) and in the presence of $K^+$, $Rb^+$, $Cs^+$, $NH_4^+$, $Mg^{2+}$ and tetraethylammonium. The overshoot increases 22-25 mV for each 10-fold increase in external [Ca]. $Ba^{2+}$ and $Sr^{2+}$ convert the graded response to one that is all-or-none. Repolarization following the peak of the response is more rapid than expected from the time constant of the resting membrane, suggesting delayed rectification. A conditioning depolarization causes inactivation of the regenerative response, and hyperpolarization depresses the response to a given stimulus current. Refractoriness persists for up to 200 msec after each response. The regenerative response of the membrane is consistent with a transient increase in conductance to $Ca^{2+}$.

The graded "calcium response" of _Paramecium_ resembles those of arthropod muscle fibers (11,12, 27,28) in which the surface membrane exhibits membrane potenital-dependent permeability increases to the bivalent alkali-earth cations. Similarities also exist with the calcium response produced by nerve terminals after the Na system is suppressed with tetradotoxin and delayed rectification is reduced with TEA (14). The response in _Paramecium_ exhibits certain properties characteristic of conventional action potentials. These include regenerative depolarization, refractoriness, and active repolarization. The calcium response in _Paramecium_ appears to perform two functions: first, it amplifies the depolarizing receptor potential which spreads electrotonically from the region of transducer membrane at the anterior surface of the cell (8,22). The final depolarization is graded since the receptor current and regenerative response are both graded. Second, the increased Ca permeability evoked by the electrotonic spread of depolarizing receptor current leads to a distributed influx of $Ca^{++}$ which activates the mechanism for ciliary reversal (3,7).

_Electro-mechanical Coupling_ (6,7,20). Several lines of evidence suggest that the mechanical response (i.e., reversal) of the cilia is coupled to membrane depolarization by the influx of $Ca^{2+}$ through the surface membrane which produces the regenerative calcium response (3). In this hypothesis the influx of calcium in response to stimulation plays a role similar to that of the influx of $Ca^{2+}$ in the electro-mechanical coupling of smooth muscle and amphibian cardiac muscle. The proposed coupling sequence is: membrane depolarization → increased calcium conductance → $Ca^{++}$ influx increased intracellular [Ca] → ciliary reversal. According to this scheme the injection of depolarizing current into the cell will produce reversal only if it produces a sufficient net influx of calcium through the surface membrane.

Machemer used a high speed cine system in our laboratory to correlate more quantitatively the mechanical responses of beating cilia with the electric behavior of the cell membrane. By simultaneous filming and electrical recording and stimulation at low and moderate depolarizations the latency and duration of ciliary reversal were quantitatively related to the amplitude of the membrane calcium response. At stimulus intensities greater than required to produce a maximal calcium response the duration of ciliary reversal became progressively shorter and the latent period progressively longer. With sufficiently positive intracellular potentials, reversal of the cilia was completely suppressed until after the end of the potential pulse. This is taken to be due to the lack of sufficient electrochemical gradient for $Ca^{2+}$ when Vm approaches the calcium equilibrium potential.

When the stimulus pulse was sufficiently strong to suppress ciliary reversal, the cilia always reversed their beat after the end of the current pulse. This we term the "off" response; it is explained as follows: The calcium conductance ($g_{Ca}$), which has remained fully or partially turned on while the membrane potential is at or beyond the suppression level, shuts down after the end of the current pulse with a time course that lags behind the repolarization of the membrane. During this period the electrochemical driving force on $Ca^{2+}$ rapidly increases, while $g_{Ca}$ is still high. This produces, at the end of the high-potential pulse, an inward calcium current, causing the intracellular calcium concentration to undergo a transient increase. This in turn produces a brief period of reversed ciliary beating. The suppression phenomenon and the "off" response of cilia at the end of strong stimulation by outward current is reminiscent of suppression and "off" responses demonstrated by Katz and Miledi (13) for the Ca-dependent release of the neural transmitter substance from presynaptic nerve terminals.

These data support the hypothesis that the mechanical response (i.e., ciliary reversal) is coupled to membrane depolarization by the same conductance-limited influx of calcium ions which produces the graded calcium response of the membrane. Calculations (3) show that a calcium response of one millivolt or more can raise the intraciliary concentration of calcium ions by $10^{-6}$M or more. The calcium response of the membrane is, therefore, adequate, in theory, to activate ciliary reversal, assuming the living cilium has the same sensitivity as that of the Triton-extracted system (24). Conversely, our demonstration that ciliary reversal is associated with calcium responses capable of producing increments in calcium concentration in this range lends physiological significance to the reported calcium sensitivity of Tritonated *paramecia*.

The role of a membrane calcium response in the regulation of ciliary motility in the protozoa as indicated by these findings resembles the role of membrane-regulated calcium fluxes in the control of cellular functions in the Metazoa, as, for example, muscle contraction, transmitter release, and the conversion of the inactive enzyme phosphorylase *b* to active phosphorylase *a*. Similarities between the calcium gating properties of the membranes of *Paramecium* and nerve terminals in the control of ciliary activity and synaptic transmitter release are especially striking.

*Membrane Mutants* (18,19,26). Electrophysiological studies were done on a mutant of *P. aurelia*, first selected for and recognized through its altered locomotor behavior (15). This mutant was named "Pawn" because it fails to swim in reverse, showing forward locomotion only, even when subjected to those stimuli which invariably lead to reversed swimming (ciliary reversal) in the wild type.

Pulses of long duration (400 msec) were passed in both outward (depolarizing) and inward (hyperpolarizing) directions in wild type and Pawn. Wild type *P. aurelia* behave electrically like wild type *P. caudatum*, showing graded regenerative depolarizing transients in response to depolarization, and electrotonic potentials (with some delayed anomalous rectification) in response to hyperpolarizing current. In contrast, Pawn gave no sign of electric excitation, never showing an inflection of the upstroke of the RC component or an oscillation on the plateau. Barium renders the wild type membrane highly sensitive to depolarization (21) so that depolarizing current as low as $10^{-11}$ amperes produced all-or-none action potentials. In contrast, Pawn remained completely inexcitable in $Ba^{2+}$.

The mutant membrane of Pawn behaves like wild type in essentially all measured electrical characteristics except one: it gives no indication of a regenerative response to depolarization. The concomitant loss of both ciliary reversal (controlled by $Ca^{2+}$ influx) and regenerative depolarization (produced by $Ca^{2+}$ current) supports the conclusion that Pawn is deficient in the mechanism which allows $Ca^{2+}$ ($Ba^{2+}$, $Sr^{2+}$) to enter the cell through the membrane in response to a depolarization.

The modified membrane function (i.e., loss of $Ca^{2+}$ activation) gains special significance from the evidence that Pawn is the result of a mutation of a single gene (16). Thus, it appears that the mechanism for calcium activation was lost due to depletion or modification of a single (or small group of) gene product(s). This may be a significant step toward the eventual isolation and characterization of the molecular components of the membrane directly involved in electrical excitation.

## REFERENCES

1. Adler, J.: Chemoreceptors in bacteria. Science: 166: 1588-1597, 1969.

2. Diamond, J., and Wright, E.: Biological mem-membranes: The physical basis of ion and non-nonelectrolyte selectivity Annual Rev Physiol 31: 581-646, 1969.

3. Eckert, R.: Bioelectric control of ciliary activity Science 176: 473:481, 1972 a.

4. Eckert, R.: Electric mechanisms controlling locomotion in the ciliated protozoa. In Behavior of Microorganisms, Plenum Press, New York, 1972b.

5. Eckert, R., and Naitoh, Y.: Graded calcium spikes in *Paramecium*. In 3rd Intern Biophys Congr Intern Union Pure and Applied Biophys, (Abstract), Cambridge, Massachuestts, 1969, p. 257.

6. Eckert, R., and Naitoh, Y.: Passive electrical properties of *Paramecium* and problems of ciliary coordination. J Gen Physiol 55: 467-483, 1970.

7. Eckert, R., and Naitoh, Y.: Bioelectric control of locomotion in the ciliates. J Protozool 19: 237-243, 1972.

8. Eckert, R., Naitoh, Y., and Friedman, K.: Calcium in bioelectric and motor functions of *Paramecium*. J Exp Biol 56: 683-694, 1972.

9. Eckert, R., Naitoh, Y., and Machemer, H.: Cal-Calcium in the bioelectric and motor functions of *Paramecium*. In Calcium in Biological Systems, Duncan, C.J. (Ed.), Cambridge, Massa-Massachusetts, 1976.

10. Eisenman, G.: Cation-selective glass electrodes and their mode of operation. Biophys J 2: 259-323, 1962.

11. Fatt, P., and Ginsburg,B.L.: The ionic requirements for the production of action potentials in crustacean muscle fibers. J Physiol 142: 516-543, 1958.

12. Fatt, P., and Katz, B.: The electrical properties of crustacean muscle fibers. J Physiol 120: 171-204, 1953.

13. Katz, B., and Miledi, R.: A study of synaptic transmission in the absence of nerve impulses. J Physiol 192: 406-436, 1967.

14. Katz. B., and Miledi, R.: The effect of prolonged depolarization on synaptic transfer in the stellate ganglion of the squid. J Physiol 216: 503-512, 1969.

15. Kung, C.: Genic mutants with altered system of excitation in *Paramecium aurelia*. I. Phenotypes of the behavioral mutants. Z Vergl Physiol 71: 142-162, 1971a.

16. Kung, C.: Genic mutants with altered system of excitation in *Paramecium aurelia*. II Mutagenesis, screening, and genetic analysis of the mutants. Genetics 69: 29, 1971 b.

17. Kung, C.: Behavioral mutants in *Paramecium aurelia*. In *Recent Progress in Microbiology*, Vol. 10, 1971 c.

18. Kung, C., Chang, S.Y., Satow, Y., Van Houten, J., and Hansma, H.: Genetic dissection of behavior in *Paramecium*. Science 188: 898, 1975.

19. Kung, C., and Eckert, R.: Genetic modification of electric properties in an excitable membrane. Proc Nat Acad Sci USA 69: 93-97, 1972.

20. Machemer, H., and Eckert, R.: Electrophysiological control of reversed ciliary beating in beating in *Paramecium*. J Gen Physiol 61: 572-587, 1973.

21. Naitoh, Y., and Eckert, R.: Electrical properties of *Paramecium caudatum*: modification by bound and free cations. Z Vergl Physiol 61: 427-452, 1968.

22. Naitoh, Y., and Eckert, R.: Ionic mechanisms controlling behavioral responses of *Paramecium* to mechanical stimulation. Science 164: 963-965, 1969.

23. Naitoh, Y., and Eckert, R.: The control of ciliary activity in protoza. In *Cilia and Flagella*, Sleigh, M.A. (Ed.), Academic Press, New York, 1974.

24. Naitoh, Y., Eckert, R., and Friedman, K.: A A regenerative calcium response in *Paramecium*. J Exp Biol 56: 677-681, 1972.

25. Naitoh, Y., and Kaneko, H.: ATP-Mg-reactivated Triton-extracted models of *Paramecium*: modification of ciliary movement by calcium ions. Science 176: 523-524, 1972.

26. Satow, Y., and Kung, C.: Mutants with reduced Ca activation in *Paramecium aurelia*. J Membrane Biol 28: 277-294, 1976.

27. Werman, R., McCann, F.V., and Grundfest, H.: Graded and all-or-none electrogenesis in arthropod muscle. I. The effects of alkaliearth cations on the neuromuscular system of *Romelia microterpa*. J Gen Physiol 44: 979-996, 1961.

28. Werman, R., and Grundfest, H.: Graded and all-or-none electrogenesis in arthropod muscle. II. The effects of alkali-earth and onium ions on lobster muscle fibers. J Gen Physiol 44: 997-1027, 1961.

# *Learned Behavior in Diptera and the Use of Mutations*

Jerry Hirsch

I feel more comfortable thinking about our subject under the French title of this conference: "Mutations: Biologie et Société, Une réfléxion pluridisciplinaire" than under its English title: "A Multidisciplinary Approach to Mutations--Biology and Society." Thus, from the point of view that makes sense to me, my "réfléxion" about the subject of this morning's round table is easy to express: the consequences of mutation are individual differences in behavior and on the average in any phenotypes that we might measure.

The distinguished American behavioral biologist, Donald Griffin of Rockefeller University, renowned for his superb analysis of echolocation by bats, has spent many years studying orientation, navigation, and migration in birds. There is a section omitted by Darwin from the Instinct chapter of the Origin of Species (1859), but later published by Romanes as an appendix to Mental Evolution in Animals (19). In that discussion Darwin, who was a geologist as well as a biologist, explained quite simply how an animal species might come to migrate over long distances. Darwin pointed out that, if long ago a species did something like build its nest in one part of its habitat and something else, like forage for food, in another part and slowly the two places moved away from one another as the land masses, on which they were situated, drifted apart, then after sufficient geologic time had elapsed for the intervening distance to grow large, the species in question would have become migratory through natural selection of those completing the journey every generation. I understand that Seymour Benzer presented a colloquium on his behavioral mutants in *Drosophila* to Rockefeller University. In the course of his talk there he turned and said to Donald Griffin, "If I were studying bird migration, I should look for a mutation."

I doubt that Benzer was alluding to the Darwin account. Benzer's and Darwin's ways of thinking about the problem of navigation represent the two classic approaches to the phenomena of biology: mechanism and evolution. The approaches are not mutually exclusive. Both should be considered in any complete analysis, but the advocates of one rarely go on to consider the other.

Today I consider the state of our substantive knowledge about Dipteran learning and some poorly appreciated methodological considerations fundamental to the study of that very elusive phenomenon, as well as the important role mutation could play in further analysis when reliable phenotypic measures are developed for an appropriate species. The goal of the approach through mutations is the explanation of behavior in terms of the functioning of specific components of physiological systems and subsystems. Let me emphasize from the outset how much I value detailed information about the gene composition of the cell systems of the organs implicated in particular behaviors. There are two extreme views about the value of such knowledge. They can be characterized as isomorphism versus independence. If individual differences (IDs) in behavior could be explained by genetic influences uniformly effective over varying environments, isomorphism would be the correct view. If they could be explained by environmental influences uniformly effective over different genotypes, independence would be the correct view. Neither view can be justified by our knowledge about Diptera. That is why it is now necessary to become much more explicit about the kind of knowledge to be expected from behavior genetic research.

It is important that we remember that the result of a successful genetic analysis is finding genes. That has been the great achievement of the *Drosophila* work. The outcome of successful behavioral analysis is the identification and description of patterns of behavior, of the components of each pattern, and understanding behavior in an evolutionary ecological context. That has been the great achievement of ethology.

Médioni (14) described a masterful use of mutation for behavior study in his elegant analysis of *Drosophila* phototropism. There is a fourth chromosome locus in *D. melanogaster*, with an allele yielding flies without eyes, called eyeless. About one out of every hundred flies carrying the mutant allele, however, develops a single eye only, with the other eye remaining undeveloped. Médioni was able to use the one-eyed flies as a control to demonstrate the relationship between directional photic stimulation and the phototropic tendencies of *D. melanogaster*. Light-adapted flies with two normal eyes are attracted by and move toward a light source. They are photopositive. When such flies have one eye covered by an opaque paint, they also move toward the light source, but do so by means of awkward circus movements. They spin around, always turning in the direction of the sighted eye. Thus, they approach the light by means of a series of spiral turns. In order to control for possible effects of tactile stimulation from the veneer of paint cover, Médioni employed the naturally occuring but rare one-eyed flies carrying the eyeless allele. They also manifest positive phototropism by spiralling toward the light, continually turning in the direction of their single functional eye. In this way the use of the mutation provided an essential control for validating the interpretation of a frequent experimental manipulation. It is equally important to note the manner in which that mutation was employed: the norm of developmental reaction for the eyeless gene included rare cases with a unique eye.

Figure 1. Vertical ten-unit plastic maze facing a fluorescent tube. Squads of files introduced in the vial are collected from the vials at right. They are attracted through the maze by the odor of food and by light. Small trap-like funnels, having a larger opening continuous with the alley surfaces and a small one debouching in midair, discourage backward movement in the maze (Hirsch 1959).

My multiple unit mass screening maze (see Figure 1) has made it relatively easy to measure IDs and to compare distributions from different populations with respect to geotropism. Large numbers of individuals can be tested together and for geotropism we can record distributions of scores ranging from consistently positive through neutral to consistently negative.

Kessler and Kraemer (9) used two laboratory populations--the Canton-S(C) and Oregon RC(0)--in both of which they placed apricot eye ($w^a$) and white eye (w)--mutations at two separate X-chromosome loci. These mutations significantly influence geotropic performance in the maze. One of the findings by these investigators is of particular interest for our discussion today. Each locus was made heterozygous with its wild-type allele and the two mutations were used together in the alternative cis and trans arrangements in both the Canton and the Oregon genetic backgrounds. In both genetic backgrounds the shift from the cis to the trans arrangement had a significant effect on geotropic score, but the phenotypic effects were in the opposite direction in the two genetic backgrounds. The shift from cis to trans in Canton increased the score but in Oregon decreased it.

## GENETIC ANALYSIS AND CONDITIONING IN A MODEL SPECIES

Considerable interest was aroused when several reports on *Drosophila melanogaster* appeared by Benzer and collaborators extolling the advantages of mutant analysis both of tropistic behavior (1) and of what has been mistaken for conditioning (18,17,4). Unfortunately, careful study reveals both the genetic and behavioral situations to be very obscure. The genetic discussions advocated using "an inbred strain and isolat(ing) mutants" as superior to starting "with a genetically heterogeneous pool. . .[and doing] progressive selection" (1). Also, it was stated that "Many flies of identical genotype are readily produced, so that behavioral measurements can be made on populations rather than individuals, yielding instant statistics" (18). But it now appears that the reported observations were not made on "flies of identical genotype," because the Canton Special (C-S) strain being used has not been maintained by any of the usual rigorous *Drosophila* inbreeding regimens (Benzer, private communications 20 September and 26 October 1976), which involve mating individuals of known close family relationships. (C-S) is a laboratory maintained, freemating, wild-type strain and, as Wright pointed out long ago, "matings more remote than between first cousins are of virtually no significance as inbreeding" (23). Furthermore, Clayton and Paietta (2) have shown that _D. melanogaster_, maintained in the laboratory over 47 years--the Oregon-R line--can have more trait relevant (additive) genetic heterogeneity than a conspecific wild population from a natural habitat. Certainly, irrespective of inbreeding status, using "a potent mutagen" (1) will produce mutations throughout the genome and thus introduce additional genetic heterogeneity (heterozygosity). Therefore, until unambiguous evidence on the genetic status of (C-S) is forthcoming, the flies described in those reports cannot be assumed to be "of identical genotype." On the other hand, Médioni and Vaysse (15) have sibmated the Basel-06 strain for 70 generations and waited till generation 40 of inbreeding to begin their study of conditioned behavior (Médioni, private communication 2 November 1976).

For almost half a century, reports have occasionally appeared claiming success in demonstrating conditioning in one or another species of the fastbreeding and prolific Diptera. The great attraction has been the hope of studying a convenient system in which can be manipulated experimentally both the dimensions of the developmental experiences involved when an individual learns and the genetic correlates of reliably measured IDs in learning. Even though the Benzer group discussed habituation and cited the studies by Manning and by Yeatman and Hirsch, somehow it has not been appreciated that their claim of successful conditioning is invalidated by their own control observations, which fail to support, and in fact, contradict, their interpretation. The analyses in those earlier studies, moreover, remain both valid and pertinent.

> Manning (1967) has made the valuable suggestion . . .[which is] the test necessary to distinguish between the alternative interpretations

> of habituation and conditioning. Such a test would involve running for a second trial those flies choosing the odour on the first trial. A conditioning interpretation predicts that those flies initially choosing an odour would choose it again on the second trial. If only habituation. . . were involved the choice on the second trial might be random" (24).

In none of the recent studies did *Drosophila*, which had met criterion on a test trial, perform any better on a retest than did those which had failed to meet criterion on the test. The behavior observed was unstable and the evidence was unequivocal: IDs in behavior were unreliable. Working with proboscis extension to sucrose stimulation in their attempts to condition *D. melanogaster*, Médioni and Vaysse (15) have reported failure when they used it as positive reinforcement but success when they used electric shock as negative reinforcement to inhibit the same response.

In fact, *no* report of conditioning or learning in any Dipteran has withstood the scrutiny of independent replication and detailed analyses [see Yeatman and Hirsch's review (24)] before Nelson (16), building upon the careful analyses of responses and conditions in the blowfly *Phormia regina* by Dethier, Solomon and Turner (3) and working with individual animals of this species, demonstrated classical conditioning and the reliable measurement of IDs. Heldon's has now become the first such claim for the Diptera to have been successfully replicated by an independent laboratory and analyzed in a selective breeding study validating genetic correlates for the foundation population distribution of IDs. An unconditioned proboscis extension response (UR), elicited by a 0.5-M sucrose stimulus applied to the labellar lobes, can be conditioned to a neutral stimulus of either distilled water or 1-M saline applied to the front tarsi, if control is included for sensitization, known as the central excitatory state (CES). The conditioning procedure with CES control consists of 15 trials presenting in succession three stimuli, the first two being neutral ($CS_1$ and $CS_2$) and the third unconditioned (US),[1] with temporal contiguity but not overlap between $CS_1$ and $CS_2$ (each presented for 4 seconds), but partial overlap between $CS_2$ and US (presented during the fourth and last second of $CS_2$), and a 10-minute intertrial interval (ITI). Two neutral stimuli, rather than one, permit $CS_1$ to drain off responses resulting from US-produced CES and $CS_2$ to measure the strength of the conditioned association. The 10-minute ITI permits the CES to wane. Flies, deprived of food and water for three days from eclosion, were mounted by depressing their partially opened wings into tackiwax-coated paraffin cylinders on wooden applicator sticks. Over 15 trials, the proportion of flies responding with proboscis extension increased steadily to $CS_2$ but not to $CS_1$. The first seven trials constituted the training period, and the last eight, the test in which IDs were measured by three categories of conditioned response (CR) score: good (6, 7 or 8), fair (3, 4, or 5), and poor (0, 1, or 2). In addition to the foregoing, Nelson performed six control experiments.

Our laboratory is pleased and proud to report not only that the *P. regina* observations are replicable but also that they lead to many analytic possibilities. In what follows, I describe first the nature of our primary observations and then the subsequent analyses: those already completed, still in progress, or projected for the future. We have now corroborated Nelson's basic results in an independent replication (11). Furthermore, both studies agreed in their distributions of IDs when flies were graded as 'good,' 'fair,' and 'poor' learners according to their number of conditioned responses over the final eight trials. This suggested the possibility of selection and further behavior-genetic analysis. Bidirectional selective breeding was, therefore, undertaken to produce lines of bright (B) and dull (D) flies. The criteria for bidirectional selection were 'good' and 'poor' performance in our replication of Nelson. A sample of 39 flies was trained, graded, and segregated into three groups of 14 good, 10 fair, and 15 poor learners. The B and D lines were established with the 14 good and 15 poor learners, respectively, along with a control line of untrained flies from another sample. After training, individuals were carefully removed from applicator sticks (to minimize wing injury) and mated. Beginning with generation 3 in both lines, the sexes were segregated within eight hours after eclosion to obtain virgin females for selection. The lines were maintained at 27°C on a 15-hr light/9-hr dark schedule.

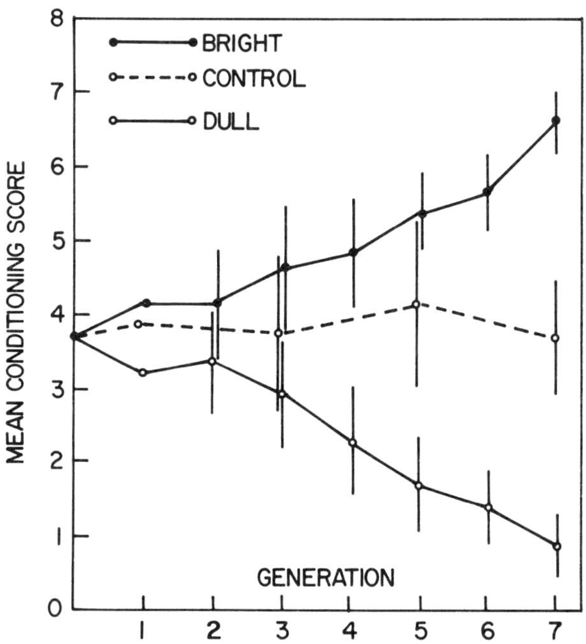


Figure 2. The mean conditioning scores (average conditioned responses to $CS_2$ on trials 8-15) and 95 percent confidence intervals for each generation of selection. (Some confidence intervals have been slightly displaced for clarity). The selected lines were tested every generation and the control on generations 1, 3, 5, and 7 (McCauley and Hirsch 1977).

In this study, the order of presentation of $CS_1$ and $CS_2$ (water and saline) was reversed on alternate generations of selection. Sample sizes varied from 20 to 39 over the three populations and the eight generations. Figure 2 compares results for both selected lines and the unselected control. By generation 7, not only did the selected lines differ significantly from one another, but both distributions differed significantly from the control. There were no sex differences. The response to bidirectional selection was rapid, especially since female virginity was uncertain before generation 3. This suggests considerable genetic variance and (together with observed phenotypic variance) that the behavior is not an important fitness trait.

Building on the foundation provided by our validation of the procedure, of the behavior, and of the existence of genetic correlates for the observed IDs, our laboratory is now engaged in the most thorough behavior-genetic analysis that is presently possible using conditioning in the metazoa.

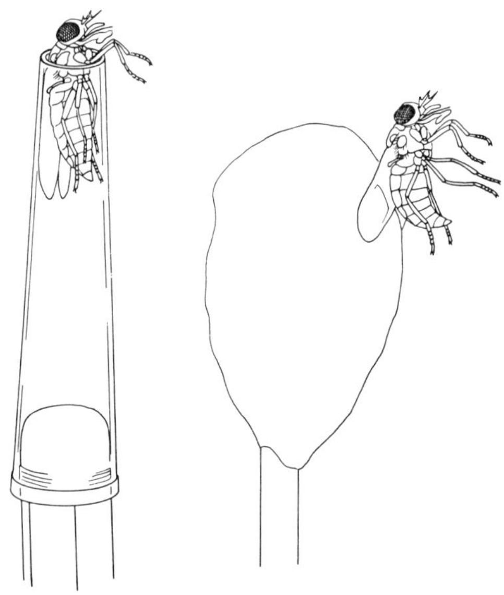

Figure 3. Comparison of fly restraining techniques: wings pressed into tackiwax (Nelson 1971) versus micropipet tip body sheath (McGuire and Hirsch 1977).

We are using the Nelson paradigm with the following modifications (13). Each fly is introduced into a disposable plastic micropipet tip from which the end has been cut off leaving a hole just large enough for the fly's head and forelegs to protrude. The fly is eased forward into position (head and forelegs protruding) with the soft bristles of a paintbrush (see Figure 3). This arrangement is mounted on a wooden rod, and 10 such mountings are placed in a holding rack and suspended horizontally about 36 cm above the surface of a table at which E is seated. This improved restraining technique has the advantage of preserving the wings (used later in mating) and appears to be less stressful. Two test tubes of 100-mm depth and 13-mm diameter containing $CS_1$ and $CS_2$ (distilled $H_2O$ and 1.0 NaCl), respectively, are taped together and raised in E's left hand so that the fly's front tarsi are immersed in $CS_1$ for 4 seconds and then immediately in $CS_2$ for the next 4 seconds. During the fourth and last second of $CS_2$ presentation, the US (a drop of 0.5-M sucrose) is touched to the labellum with a capillary tube by E's right hand. Whereas in previous work a fly was moved to and from both conditioned stimuli and the resting position, now an immobile fly can be stimulated, thus eliminating several irrelevant sources of excitation. In one 2.5 hour session 10 flies are trained and tested over 15 trials with a 10-minute ITI. The 4-seconds stimulus intervals are timed with a metronome and the 10-minute ITI with a stopwatch.

The CR is proboscis extension during the first 3 seconds of $CS_2$. Responses to $CS_1$ afford some (undoubtedly not complete) control for CES effects and those to US represent URs. During the test, any individual failing to emit at least five responses to US is discarded. As in previous studies, when the tarsi are immersed in water, any fly, that responds with full proboscis extension and labellar lobes wide open, is allowed to drink in order to compensate for possible dehydration during the experiment.

## BIDIRECTIONAL AND STABILIZING SELECTION

Using the modified Nelson paradigm, a sample (N=53) from the same freemating wild-type foundation population as in the previous study was trained, tested, and classified into a distribution of reliably measured IDs, from which bidirectional selective breeding was started. As before, "good" flies (eight pairs) with the highest scores were isolated to found the "bright" line (B), but now "poor" flies (eight pairs), isolated to found the "dull" line (D), have been chosen to exclude individuals scoring zero, i.e., stabilizing selection has been used for D, because flies might never respond for a variety of reasons unrelated to innate conditionability (pathology, trauma, etc.). All females have been virgin. A control line (C) was started with eight pairs from an untested sample and continued in that way for six generations, since when it has been maintained with 20 pairs. Over the generations, sample sizes ranged from a minimum of 29 to a maximum of 79.

Enough flies have been tested in every selected generation (with exceptions noted below) to find eight pairs qualifying as "good" in B and "poor" in D to breed B and D, respectively. Of course, selection intensity decreased with the progress of selection and the concomitant increase in the proportion qualified for breeding. Every third generation C was also tested. Again, the order of presentation of the two stimuli was reversed on alternate generations of selection, being $CS_1(H_2O)$--$CS_2(NaCl)$ for even-numbered and $CS_1(NaCl)$--$CS_2(H_2O)$ for odd-numbered generations. Selection was continuous over the first 12 generations (except generation 2 of D was unselected) then intermittent over the next 15, having been relaxed for both lines at generations 13, 15, 16, 17, 20, 21, 22, 25, and 26.

Table 1 presents the distribution of IDs for the foundation population in comparison with the distributions from three other studies. Despite differences in methods and populations the results are evidently quite similar. Figure 4 shows for B

TABLE 1. Conditioning Score Proportions

Comparison of ID distributions. The first two studies used the Nelson tackiwax mounting but different populations. The second and fourth used the same population but different mounting and stimulus arrangements. The third and fourth used the same mounting and stimulating arrangements, but the third tested three populations--the Dethier population (DT) used by Nelson, the Illinois Entomology population (E) used by McCauley and Hirsch and by McGuire and Hirsch, and the derivative control line (C) from McGuire and Hirsch. N = number of subjects.

| | | $CS_1 - CS_2$ ($H_2O$ - NaCl) | | | $CS_1 - CS_2$ (NaCL - $H_2O$) | | | |
|---|---|---|---|---|---|---|---|---|
| | | Good | Fair | Poor | Good | Fair | Poor | N |
| Nelson (1971) | | .39 | .24 | .37 | .31 | .18 | .51 | 200 |
| McCauley & Hirsch (1977) | | .26 | .30 | .44 | .26 | .26 | .48 | 39 |
| | DT | .32 | .26 | .41 | | | | 99 |
| Jackson | E | .23 | .31 | .44 | | | | 89 |
| (1976) | C | .03 | .41 | .55 | | | | 29 |
| McGuire & Hirsch (1977) | | .27 | .32 | .41 | .24 | .34 | .42 | 53 |

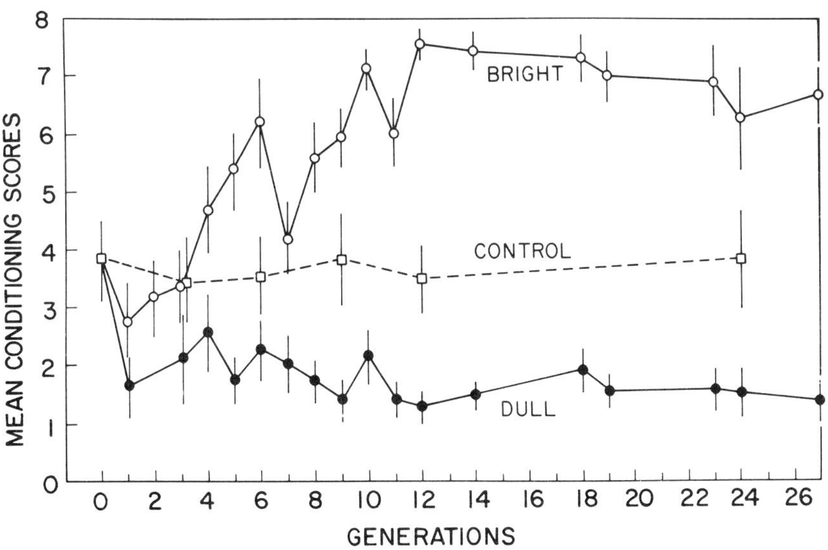


Figure 4. See legend on page 192.

Figure 4. Mean conditioning scores (average of $CR_S$ to $CS_2$ on trials 8-15) and 95 confidence intervals for B, D, and control lines over 27 generations of selection (see text for description). Some confidence intervals have been slightly displaced for clarity (McGuire and Hirsch 1977).

and D the response to selection in comparison with C over 27 generations. The initial fluctuations remained parallel up to generation 5 when divergence began. The separation of B and D from C became statistically significant at generation 8. The marked drop by B at generation 7 was correlated with an unfortunate release of insecticides into the building atmosphere.

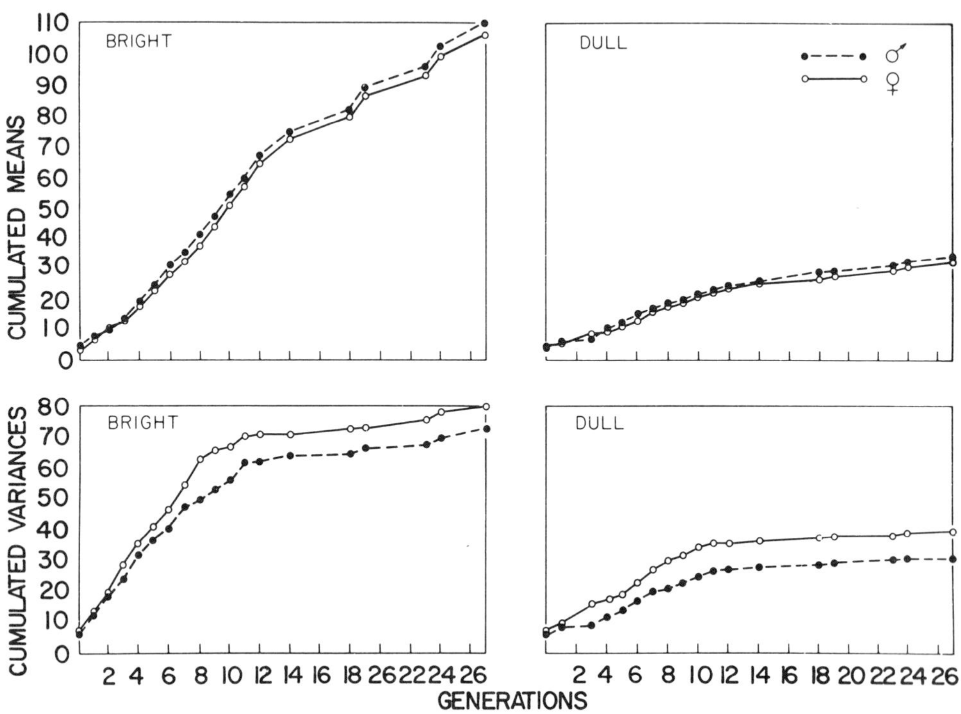


Figure 5. Cumulated means and variances of B and D lines over generations showing consistently greater female variance (Mcguire and Hirsch 1977).

Figure 5 shows cumulated means and variances for the sexes separately. Females have a larger variance than males in both lines but have a lower mean only in B. The success of stabilizing selection can be seen in Figure 6, which compares the proportion of D individuals never responding in the present and previous studies.

There are several uniquely important features of these observations. They represent the second corroboration of what is now the only independently replicated evidence for conditioning throughout the Diptera. The behavioral changes have been observed in reliably measured and identifiable individuals. The response to selection validates the existence of genetic correlates for the foundation population distribution of IDs and begins the genetic analysis of Phormia conditionability. Hybridization and backcross analyses are in progress, in which McGuire (12) now has evidence for

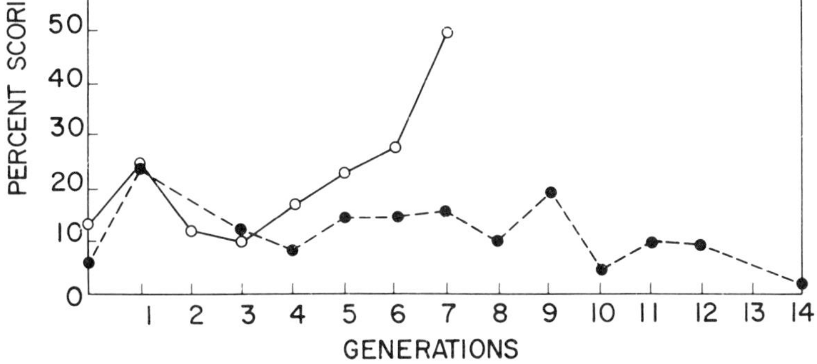


Figure 6. Response to divergent and stabilizing selection by D lines in two studies (McCauley and Hirsch 1977, McGuire and Hirsch 1977), showing importance of including or excluding zero score in selection criterion.

partial directional dominance of brightness. It should be noted that these two successful *P. regina* selective breeding studies have been done with a laboratory strain maintained well over 20 years with no less inbreeding than the Benzer *D. melanogaster* (C-S) strain. Certainly the individuals in these strains were *not* of identical genotype.

In the recent reports of conditioning in *Drosophila* there was no evidence for IDs or stability over time. Quinn *et al*. (18) reported finding no "intelligent subset." It was precisely to the question of stability in *Phormia* of the IDs in conditioning that Jackson has addressed himself (8). Though he takes strong exception to the use of the designation "intelligent subset," Jackson has shown not only that IDs can be measured reliably but that, when stored at 1.8$^{o}$C, the behavior of the individuals is stable over time. Flies, that on one day achieved high and low conditioning scores, respectively, maintained their relative differences in performance when retested at the same time 24 hours later. Wilson obtained excellent evidence for extinction (N=59), extinction followed by spontaneous recovery (N=20), and extinction followed by reconditioning and extinction again (N=10) in selected generations 23 and 24 of the B line (22).

Even the simplest conditioned behavior is complex, consisting of US-UR and CS-CR relations. In the present instance, at the very least, we are dealing with US-UR, $CS_2$-CR, US-CES and ITI-CES relations, in addition to IDs in each. Only detailed behavior-genetic analyses might unravel the strands of this complexity; and such studies, now in progress, are analyzing separately and then conjointly the distinguishable phenotypic components of this complex, i.e., lines with different degrees of expression of CES are being bred by selection from the foundation population. Their analysis can elucidate the role of CES in conditioning. Also, the expression of CES in the B and D lines can now be studied in a complementary behavioral analysis. So Nelson's cryptic suggestion that her "classical conditioning. . .is essentially an extension of the CES phenomenon" thus becomes amenable to detailed experimental behavior-genetic analysis. Is conditioning in *Phormia* "merely" CES or are there separate "states," with the habit developing in a context of optimum excitation? As the behavior measurements are refined and components are identified and their analysis is also refined, the justification for mutant analysis will become more compelling.

The observation made by Professor Manning five years ago remains pertinent: "So far as the qualitative nature of gene action on behavior is concerned, our concepts have advanced little in the past decade. We have more examples, but they offer little scope for generalization" (10). Unfortunately, it is not only the genetics, race, and IQ literature (5,7) which fails to inspire the confidence needed to undertake the kind of detailed genetic analysis for which mutations can be so effectively employed. Until we institutionalize the importance of critical replication analysis by independent investigators (21), too many of our most cited examples will remain of doubtful credibility and, therefore, uninspiring. As Schein (20) has made so clear, we now know the value of mutations for dissecting "a complex physiological process. . . In the very least, the number of genes directly affecting the process may be counted and a clearer idea of its true complexity obtained. The mutants may direct further physiological dissection, and at best, may even assist in molecular characterization." Before such a program can be implemented in the behavior study field, however, we must cleanse its Augean stables. We need what is still too rare: dependable observations in appropriate species. The advantages of *Drosophila* for genetic analysis are too well known to warrant repetition, but ease of genetic analysis is no substitute for reliability and validity of behavioral measurement.

## REFERENCES

1. Benzer, S.: Behavioral mutants of Drosophila isolated by counterdistribution. Proc Natl Acad Sci USA 58: 1112-1119, 1967.

2. Clayton, D.L., and Pietta, J.V.: Selection for circadian eclosion time in Drosphila melanogaster. Science 178: 994-995, 1972.

3. Dethier, V.G., Solomon, R.L., and Turner, L.H.: Sensory input and central excitation and inhibition in the blowfly. J Comp Physiol Psychol 60: 303-313, 1965.

4. Dudai, Y., Jan, Y.N., Byers, D., Quinn, W.G., and Benzer, S.: Dunce, a mutant of Drosophila deficient in learning. Proc Natl Acad Sci USA 73: 1684-1688, 1976.

5. Gillie, O.: Crucial data was faked by eminent psychologist. The Sunday Times, London, pp. 1 - 2, 24 October 1976.

6. Hirsch, J.: Studies in experimental behavior genetics: II. Individual differences in geotaxis as a function of chromosome variations in synthesized Drosophila populations. J Comp Physiol Psychol 52: 304-308, 1959.

7. Hirsch, J.: Jensenism: The bankruptcy of "science" without scholarship. U.S. Congr Rec 122 73:E2671-2; 74:E2693-5; 75:E2703-5, E2716-8, E2721-2, 1976. (Originally published in Educational Theory 25: 3-27, 1975).

8. Jackson, D.M.: Individual learning (classical conditioning) stability in the blowfly (Phormia regina). M.A. Thesis, University of Illinois, Urbana-Champaign, 1976.

9. Kessler, S., and Kraemer,H.: Gene substitutions and geotaxis in Drosophilia melanogaster. J Comp Physiol Psychol 89: 274-278, 1975.

10. Manning, A.: Discussion. In Ehrman, L., Omenn, G.S., and Caspari, E. (Eds.): Genetic, Environment, and Behavior: Implications for Educational Policy. Academic Press, New York, 1972, p. 48.

11. McCauley, L.A., and Hirsch, J.: Successful replication of, and selective breeding for, classical conditioning in the blowfly Phormia regina. Anim Behav 25: 784-785, 1977.

12. McGuire, T.R.: Hybridization Analysis of Classical Conditioning in the Blowfly Phormia regina. Abstract No. 50. Annual meeting Animal Behavior Society, Boulder, Colorado, 1976.

13. McGuire, T.R., and Hirsch, J.: Behavior-genetic analysis of Phormia regina: Conditioning, reliable individual differences and selection. Proc Natl Acad Sci USA 74: 5193-5197, 1977.

14. Médioni, J.: Contribution à l'étude psycho-physiologique et génétique du phototropisme d'un insecte: Drosophila melanogaster. Meigen, Strasbourg, thèse, 1961.

15. Médioni, J., and Vaysse, G.: Suppression conditionnelle d'un réflexe chez la Drosophile (Drosophila melanogaster): acquisition et extinction. Comptes rendus des séances de la Société de Biologie, 169: 1386-1391, 1975.

16. Nelson, M.C.: Classical conditioning in the blowfly (Phormia regina): Associative and excitatory factors. J Comp Physiol Psychol 77: 353-368, 1971.

17. Quinn, W.F., and Dudai, Y.: Memory phases in Drosophila. Nature 262: 576-577, 1976.

18. Quinn, W.G., Harris, W., and Benzer, S.: Conditioned behavior in Drosophila melanogaster Proc Natl Acad Sci USA 71: 708-712, 1974.

19. Romanes, G.J.: Mental Evolution in Animals. Harrison & Sons, London, 1883, p. 359.

20. Schein, S.J.: Nonbehavioral selection for Pawn, mutants of Paramecium aurelia with decreased excitability. Genetics 84: 453-468, 1976.

21. Smith, N.C.: Replication studies: A neglected aspect of psychological research. Am Psychologist 25: 970-975, 1970.

22. Wilson, S.A.: Extinction in Phormia regina: A preliminary study. University of Illinois, Urbana-Champaign, 1976.

23. Wright, S.: Systems of mating. V. General Considerations. Genetics 6: 167-178, 1921.

24. Yeatman, F., and Hirsch, J.: Attempted replication of, and selection breeding for, instrumental conditioning of Drosophila melanogaster. Anim Behav 19: 454-462, 1971.

# *Single Gene Analysis: Behavior of Children with Genetic Disorders*

Pierre Roubertoux

Michéle Carlier

Evelyne Dumont

Pierre Perruchet

The aim of this report is to present genetic analysis as one of the possible ways to approach individual differences in human behavior. Whereas previously used approaches centered on traits and were based either on the partitioning of the variation observed or on the study of pedigrees, this method uses an experimental schema where categories of the independent variable are established according to the presence or absence of one of the allele forms of a gene and the dependent variable is established by measuring both qualitative and quantitative behavior

---

[1] The means of carrying out this research were supplied by the Conseil Scientifique de l'Université René-Descartes, as a piece of interdepartmental research; by the C.N.A.M. (Service de Recherche de L'I.N.O.P.), the E.P.H.E. (Laboratoire de Psychologie Différentielle), the Université de Paris-X-Nanterre (Laboratoire de Psychologie Expérimentale et Différentielle) and the C.N.R.S. (E.R.A. no. 79). We wish to express our gratitude to Professor Royer for allowing us to install our laboratory in the Clinique de Génétique Médicale which he directs. Without him, the research described here would not have been possible.

Requests for reprints should be sent to the senior author: Service du Professeur Pierre Royer, Clinique de Génétique Médicale, Hôpital des Enfants Malades, 149 rue de Sèvres, France, 75015 Paris.

characteristics. Earlier we considered defining the category of the independent variable by recognizing mutations, the most notable of which lead to genetic disorders in children. These mutations present several characteristics which simplify genetic analysis:

- they are identifiable biochemically;
- some occur relatively frequently;
- for several of them the biochemical nature of the mutation is known;
- the consequences are often generalized to such an extent that disturbances of the cerebral function may not be ruled out; and
- finally, the knowledge of genetic diseases in animals gives opportunities for direct experimentation.

The aim of this report is not to review papers on the effect of mutations on human behavior, but rather to explain by examples a method by answering three groups of questions:

1. Which behavior disorders are associated with genetic diseases? What can these diseases teach us about the origins of pathological behavior in children?

2. Can the study of these diseases show the mechanism by which mutation affects behavior?

3. Can this approach, which was first considered for the qualitative characteristics, be extended to the quantitative characteristics?

## ASSOCIATION OF BEHAVIOR DISORDERS AND GENETIC DISEASES IN CHILDREN AND ADOLESCENTS[2]

[2] The ideas described in this paragraph have been presented during the research seminars of Professors Chiland, Diatkine, Duché, Lebovici, Pichot and Widlöcher. We have profited from the discussion which we had with the researchers on these occasions. Since the preparation of Table 1 in 1976 and its presentation in February 1977, there has come to our notice G.S. Omenn's article: Inborn errors of metabolism: clues to understanding human behavioral disorders, Behavior Genetics 6: 3 263, 1976, which presents suggestions similar to ours.

Table 1 lists the genetic diseases in children which are associated with behavior disorders whose prevalence is five times greater than in the normal population. This presentation is incomplete and, moreover, open to criticism. However, these imperfections may incite further reflection and lead to the discovery of a solution to difficulties which do not seem fundamental.

Despite the diversity of clinical syndromes taken into consideration, feeblemindedness is dominant. This may be due to the fact that geneticists often study mentally defective children in institutions. Moreover, the imprecision of the diagnosis may have to be taken into account; for example, it is often difficult to differentiate between a mentally defective and an autistic child. Finally, the term feebleminded covers children affected to a greater or lesser degree. It would, therefore, be imperative to know the distribution of global intellectual levels, subtests, and different psychometric tests. These observations illustrate the advisability of carrying out systematic psychological examinations, of using standardized tests on children with genetic diseases, and of noting in the clinical observations the instruments and numerical criteria used for the diagnosis.

In the large category entitled "personality disorders," are the clinical syndromes dealing with infantile psychosis. However, the association of genetic disease and the psychopathological syndrome is flexible. First, the association of a psychopathological syndrome and a genetic disease is made with varying frequency. It must not be forgotten that diseases where the prevalence of behavior disorders is similar to the prevalence of the same disorders in the normal population, have already been eliminated here. Then the same psychopathological syndrome is present in several genetic diseases. To take only one example among many, the behavior associated with childhood psychosis is also present with high frequency in hyperuricemia, Hartnup's disease, polydystrophic oligophrenia, several sphingolipidoses, etc.

Table 1. Association of behavior disorders with genetic diseases in childhood. Abbreviations: MR, mental retardation; PR, psychomotor retardation; IR, intellectual regression; ND, neurological disorder; E, epilepsy; PD, personality disorder; AR, autosomal recessive; ARH, autosomal recessive, heterogeneous; SLR, sex-linked recessive; H, heterogeneous; AD, autosomal dominant; M, mild; V, variable; S, severe.

| NAME OF DISEASE | NATURE OF DEFICIT | MODE OF TRANSMISSION | BEHAVIORAL MANIFESTATIONS | | | | | |
|---|---|---|---|---|---|---|---|---|
| | | | MR | PR | IR | ND | E | PD |
| AMINO ACID METABOLISM | | | | | | | | |
| Isovalericacidemia | Isovaleric Acid - Coenzyme A Dehydrogenase | AR(?) | x(M) | | | x | | |
| Argininosuccinic Aciduria | Argininosuccinase | AR | | | x | | x | |
| Citrullinuria | Argininosuccinic Acid Synthetase | (?) | | | x | | | x |
| Histidinemia | Histidase | AR | x(M) | | | | | |
| Homocystinuria | Cystathionine Synthetase | AR | x(V) | | | | | x |
| Hydroxyprolinemia | Hydroxyproline Oxidase | (?) | x | | | | | |

Table 1 (continued)

| NAME OF DISEASE | NATURE OF DEFICIT | MODE OF TRANSMISSION | BEHAVIORAL MANIFESTATIONS | | | | | |
|---|---|---|---|---|---|---|---|---|
| | | | MR | PR | IR | ND | E | PD |
| Hyperphenylalaninemia | Deficient in Phenylalanine Hydroxylase<br>Phenylalanine transaminase | ARH | x(M) | | | | x | x |
| Hyperprolinemia I | Proline Oxidase | (?) | x | | | | | |
| II | $\Delta^1$Pyrroline-5-Carboxylic Acid Dehydrogenase | (?) | x | | | | x | |
| Hypervalinemia | Valine Transminase | AR | x | | | | | |
| Phenylketonuria | Phenylalanine Hydroxylase | AR | x | | | x | x | x |
| Hyperammonemia I Russell Type | Ornithine Transcarbamylase | AR or SLR | x | | | x | x | |
| Hyperammonemia II | Carbamyl Phosphate Synthetase | AR | x | | | | x | |
| Argininemia | Arginase | AR | x | | | | x | |

Table 1 (continued)

| NAME OF DISEASE | NATURE OF DEFICIT | MODE OF TRANSMISSION | BEHAVIORAL MANIFESTATIONS | | | | | |
|---|---|---|---|---|---|---|---|---|
| | | | MR | PR | IR | ND | E | PD |
| Hartnup Disease | Keto Acid Decarboxylase | AR | x(V) | | | | x | x |
| SPHINGOLIPIDOSES | | | | | | | | |
| Gaucher Disease | | | | | | | | |
| 1. Chronic Form | Glucosyl Ceramidase | AR | | | | x | | |
| 2. Juvenile Form | Glucosyl Ceramidase | AR | | | x | x | | |
| Metachromatic Leukodystrophy | | | | | | | | |
| 1. Infantile | Arylsulfatase A | AR | | | x | x | | |
| 2. Juvenile | (?) | H | | | x | x | | x |
| Tay-Sachs Disease | β Hexosaminidase A | AR | | | x | x | | x |
| Generalized Gangliosidosis I | β Galactosidase | AR | x(S) | | | | | |
| Late Form II | (?) | AR | x(S) | | | | | |
| Neimann-Pick Disease | | | | | | | | |
| Type A | Sphingomyelinase | AR | x(S) | | | x | | |
| Type B | (?) | AR | | | x | x | | |

Table 1 (continued)

| NAME OF DISEASE | NATURE OF DEFICIT | MODE OF TRANSMISSION | BEHAVIORAL MANIFESTATIONS | | | | | |
|---|---|---|---|---|---|---|---|---|
| | | | MR | PR | IR | ND | E | PD |
| PURINE METABOLISM | | | | | | | | |
| Hyperuricemia | Hypoxanthine-guanine Phosphoribosyltrans-ferase | SLR | | x | | | | x(V) |
| MUCOLIPIDOSES | | | | | | | | |
| Type I | (?) | AR | x | | | | | |
| Type III | (?) | AR | | | | | | x(V) |
| Type IV Juvenile Sulfatidosis | Arylsulfatases A,B,C | AR | | | x | x | | |
| CARBOHYDRATE META-BOLISM | | | | | | | | |
| Galactosemia | Heterogeneous | ARH | x(V) | x | | | | |
| MUCOPOLYSACCHARI-DOSES | | | | | | | | |
| Type I - Hurler Gargoylism | $\gamma$-L-iduronidase | AR | x | | x | x | x | |

Table 1 (continued)

| NAME OF DISEASE | NATURE OF DEFICIT | MODE OF TRANSMISSION | BEHAVIORAL MANIFESTATIONS | | | | | |
|---|---|---|---|---|---|---|---|---|
| | | | MR | PR | IR | ND | E | PD |
| Type III-Sanfilippo Syndrome | (?) | AR | | x | x | x | x | x(S) |
| Type IV-Morquio Syndrome | (?) | AR | | | | | | |
| ACIDOSES | | | | | | | | |
| Congenital Lactic Acidosis | (?) | (?) | x | | | x | | |
| Methylmalonicaci-duria | Methylmalonyl CoA Mutase | ARH | | | | x | | |
| OTHER GENETIC DEFECTS | | | | | | | | |
| Acrocephalies | Heterogeneous | H | x(S) | x | | | | |
| Behr Syndrome | (?) | AR | x(S) | | | | | |
| Tuberous Sclerosis (Epiloia) | (?) | AD | x(S) | | | | | |

Table 1 (continued)

| NAME OF DISEASE | NATURE OF DEFICIT | MODE OF TRANSMISSION | BEHAVIORAL MANIFESTATIONS | | | | | |
|---|---|---|---|---|---|---|---|---|
| | | | MR | PR | IR | ND | E | PD |
| Duchenne Muscular Dystrophy | | | | | | | | |
| Rapid - I | Dyphenyloxidases (?) | AR | x(V) | | | | x | x |
| | | SLR | x(V) | | | | x | x |
| Slow - II | Diphenyloxidases (?) | AR | x(V) | | | | x | x |
| | | SLR | x(V) | | | | x | x |
| Wilson Disease | (?) | AR | | | | | | x |

Finally, different psychopathological syndromes may be associated with the same genetic disorder, for example, mental deficiency, and the hyperkinetic syndrome. A detailed examination of several examples has already been presented (15).

In the same publication, we have been led to form the following working hypothesis which may be summarized thus:

> Each psychopathological syndrome would have several possible genetic etiologies.
>
> As several genetic disorders imitate the same syndrome, it may be concluded that transmission must be heterogeneous for genetic etiologies and consequently pathologies; different genes must act independently on different biochemical targets, without the behavior disorder being modified; the study of the transmission of morphological characteristics in different species shows that such phenomena are frequent in genetics.
>
> As such psychopathological syndromes are also found independently of any genetic disorder, the hypothesis as to the heterogeneity of the etiology is reinforced; other hypotheses, more complex from the genetic point of view, may be considered but should not be proposed until simpler hypotheses (for example, monogenetic) have been disproved using adequate dependable methods.
>
> At present, nothing can be said of the variation in psychopathological manifestation except to emphasize strongly its existence.

The combination of hypotheses summarized here seems the only possible explanation of a certain number of facts which have remained inexplicable until now: 1) the accumulation of presumptions in favor of a genetic etiology of psychosis and the failure of all propositions relative to a precise means of transmission (14); 2) the seemingly contradictory results in biochemistry, which, in fact, are only the consequences of the heterogeneity of the experimental groups; 3) the evolution of the sex ratio in mental illness according to age at time of hospitalization (15).

## CONTRIBUTION OF THE SINGLE GENE ANALYSIS TO THE IDENTIFICATION OF THE EFFECTS OF MUTATION ON BEHAVIOR[3]

Not being physicians, we have not approached these problems from the point of view of cerebral dysfunction. For this, articles such as that of Brady (1) may be consulted. As psychologists and geneticists, we shall show that the simple gene analysis can solve one of the psychologist's problems, using the laws of segregation of formal genetics. Our study was carried out by our team on children with muscular dystrophy. (4) Duchenne de Boulogne's muscular dystrophies develop either rapidly (DDB I) or slowly (DDB II). Both DDB I and DDB II are recessive X-linked (80% of cases) and recessive autosomic (20% of cases) (5,9,11). It has already been shown (Table 1) that mental deficiency is prevalent to a varying degree in children with muscular dystrophy but is, in any case, more frequent than in the normal population. The relation of mental deficiency and muscular dystrophy is the subject of five hypotheses which should be examined in the light of 650 observations which we have carried out.

1. Mental deficiency can be imputed to different factors of the environment. Some blame the child's own anxiety or that of his family, or the consequences of motor impairment on intellectual development. Several facts contradict this hypothesis:

---

[3] This research, as well as that presented in the third part, was carried out with the comprehension and help of Dr. Demos, Directeur de Recherche à l'I.N.S.E.R.M., and with the support of the Comité Francais de Soutien à la Recherche contre la Myopathie (C.O.F.R.A.S.O.R.M.).

a) Behavior disorders are precocious and often appear before muscular dystrophy.

b) We have never seen any report on intellectual regression in DDB I.

c) Behavior disorders are independent of the physical impairment. If mentally deficient DDB I and nonmentally deficient DDB I patients are compared, there is no difference between the two groups as to the extent of physical impairment (for an adjusted age), for the age at which walking stops, and for the age at which the first steps were taken.

d) The strongest argument is based on the fact that in several diseases presenting the same general syndrome (as to the extent of physical impairment) as muscular dystrophy, mental deficiency is no more frequent than in the normal population. For example, two mentally deficient patients were found in a group of 139 subjects with an abiotrophy of the peripheral motoneuron [Werdnig Hoffmann's disease] (6). For a similar group of DDB I, according to the statistics of different authors, between 18 and 48 mentally deficient patients should have been found.

2. The hypothesis of a linked gene is not acceptable either, for if muscular dystrophy and mental deficiency resulted from two linked genes, nonmentally deficient children with muscular dystrophy should be found in the same family, as well as a greater number of mentally deficient children without muscular dystrophy than in the normal population. Whereas, the incidence of nonmentally deficient children with muscular dystrophy is fortunately high and only two seriously or moderately deficient cases without muscular dystrophy were found among the brothers and sisters of children with muscular dystrophy seen in the Hospital des Enfants Malades. This observation was made from a group of 198 families. A second argument is based on the fact that mental deficiency is associated with both forms of muscular dystrophy, DDB I and DDB II, and with both modes of transmission, X-linked recessive and autosomic

recessive, with a varying incidence in the different cases but always greater than in the normal population.

3. Like McKusick (11), we, therefore, opt for the hypothesis of a pleiotropic effect. As to the type of effect, it can only be a toxic factor notably creatine kinase. The correlation between the IQ and the creatine kinase level was not significant for 60 subjects. The creatine kinase level was compared in two groups of children with muscular dystrophy (DDB I), one group being mentally deficient and the other not. The difference was not significant. A similar non-significant difference relating to creatine kinase was noted when severely retarded children were compared with children of normal intelligence who had muscular dystrophy.

Two other hypotheses have been suggested:

4. A defect of neuronal migration, leading to a defective cortical integration, may be one among the effects of enzyme multipotentiality (17).

5. A microcirculatory etiology of muscular dystrophy has been supported by Demos, who has attributed the disorders of the cerebral function in children with muscular dystrophy to one of the elements of the microcirculatory syndrome, with other aspects being muscular impairment and retarded growth (3).

## SINGLE GENE ANALYSIS AND QUANTITATIVE CHARACTERISTICS

Here a difficulty arises from the psychological point of view, as the study must be carried out at a rather elemental level as regards genetic analysis, but significant enough as regards the behavior of the species. In studying the most elementary retention activities, our laboratory employed a delayed conditioning as the experimental design. The introduction of a new stimulus during the conditional stimulation, when conditioned response is established, perturbs the conditioned response. This disturbance depends on the previous situation [stability of the conditional relation, the novelty and intensity of the disturbing stimulus, and also on individual differences (12)]. Here the UCR consists of the reaction of eyelid closure to an air puff of 100 ms on the cornea. It is recorded by electromyography, using silver surface electrodes (Alvar capusles type) which are fixed with collodion in the sagittal median plane of the eyeball, approximately one-half centimeter above the eyebrow and approximately one and one-half centimeters below the inferior palpebral groove. The earth electrode is fixed two centimeters away from the external commisure. The left eye is always used. The preamplifier used has a bandwidth of 0.16 to 3,500 c/s at 3 dB. Continuous recordings with polygraph are used. The conditional stimulation (CS) consists of a sound of 1,100 hertz to 35 decibels, transmitted by earphones. The unconditional stimulus (UCS) is a puff of nitrogen arriving at 15 ± 5 millimeters from the external commisure, directed towards the cornea. The pressure of the air puff is chosen so as to bring a deflection equal to that produced by a voluntary blink. The relation between the CS and the USC is shown in Figure 1. The subjects are examined individually in a soundproof cabin of 8 m.$^3$ They are in a semireclining position. The first part of the experiment consists of the habituation phase, when the sound is presented, until the complete disappearance of the blinking orientation reactions. The second part consists of the presentation of five blocks of stimulation, each block consisting of 10 times the association of CS-US with a single presentation of CS only; these two parts of the experiment take place in darkness. After the introduction of a parasite stimulation, (here, a continuous soft light) one and one-half blocks are presented again to the subject; this is the third part. The last phase consists of the extinction of any CR. The experiment lasts approximately 30 minutes.

In the general population, during the second part of the experiment, a conditioned response is established which is reflected either by an

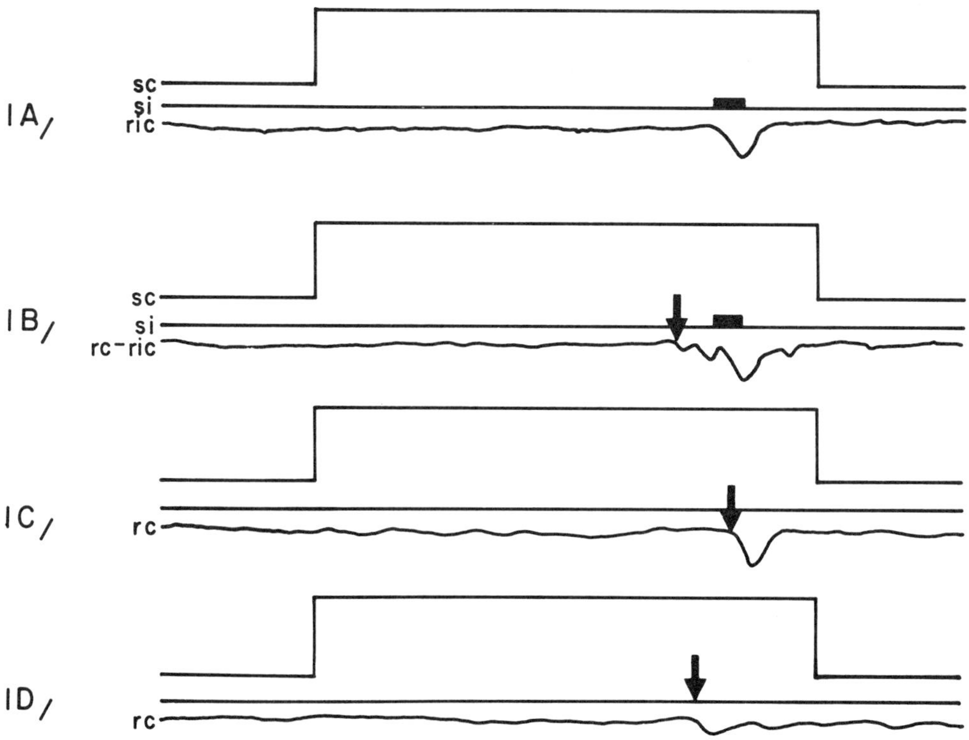


Figure 1. Example of delayed eyelid conditioning. The distance between the beginning of the sound and the puff of nitrogen is 1,600 ms.: a) unconditioned response of eyelid closure in response to puff of nitrogen; b) example of anticipated response, the arrow indicates the beginning of anticipation; c) example of conditioned response indicated by the arrow; and d) example of conditioned response after disruption by a parasite stimulus.

increase in the number of blinks and their intensity as the moment approaches for the UCS to be given (Figure 1) or a single response during a stimulus test at the exact place at which the puff of nitrogen should have been directed (Figure 1). The introduction of the parasite stimulation (soft light) disturbs the conditional relation which may be shown either by an increase in anticipation or by a reduction, or even abolition, of the conditioned response (see Figure 1). The interpretation of these results has already been discussed in detail (12). The extent of the disturbance depends, among other things, on individual differences.

The existence of an effect which may be associated with genetic differences has been verified from an experimental design described in the introduction. The categories of the independent variable consist of the modes of transmission of muscular dystrophy. Thus, there is one group of children with X-linked recessive muscular dystrophy and another group with autosomic recessive muscular dystrophy. The clinical diagnosis and the biochemical tests were made by Dr. Demos and his team. The assignation of the children to one of the categories of the dependent variable, either X-linked recessive or autosomic recessive, is made taking into account two criteria: 1) the biochemical detection of individual parents who are heterozygotes, and 2) the examination of the pedigrees. The subjects for whom we have no information relative to these two criteria together were eliminated from the sample. Thus, only the children satisfying both criteria were included. For example, a subject, whose mother is a heterozygote and who has at least one relative affected, was included in the group of X-linked recessives; a subject with both parents heterozygotes and a sister with the disease, was classed as autosomic recessive. Similarly, female patients whose mothers and fathers were both found to be heterozygotes, were put into the latter group. However, only one member of each family was included in these groups. Both groups were matched for age and scholarship. Children who were more than two years behind at school were eliminated, thus eliminating mentally deficient children. There was no difference between the groups in the degree of physical impairment. The X-linked recessive group was composed of 16 children; the autosomic recessive group consisted of 13 children: 9 boys and 4 girls. During the experiment, the group to which the subjects belonged was only known to one of the testers. The analysis of the graphs was done anonymously, the only means of identification being a number, known only to one researcher who was not concerned in the interpretation of the electromyographic recordings.

Taking as a criterion of the establishment of the conditioned response, the response appearing between 1,220 and 1,600 milliseconds, no difference was noted in the establishment of positive conditioned responses in either group of subjects. However, it seems interesting to compare the results of the two groups on the basis of time to recuperate after the introduction of a parasite stimulus (light). Several indications of recuperation were calculated. All these composite indications helped to classify the subjects in their original groups. One which took into account two disrupting elements helped to identify 90% of subjects (2). Another, taking three disruptive elements (grouped in two variables) into account, allowed the classification of 97% of subjects on the basis of discriminant functions with two variables and two groups (Figure 2).

The interpretation of these results is difficult. From the viewpoint of own field of biometric and formal genetics, several remarks can be made.

1. The hypothesis of an artifact dependent on an uncontrolled variable should not be considered here, at least in the case of age, mental deficiency, or the extent of impairment which were

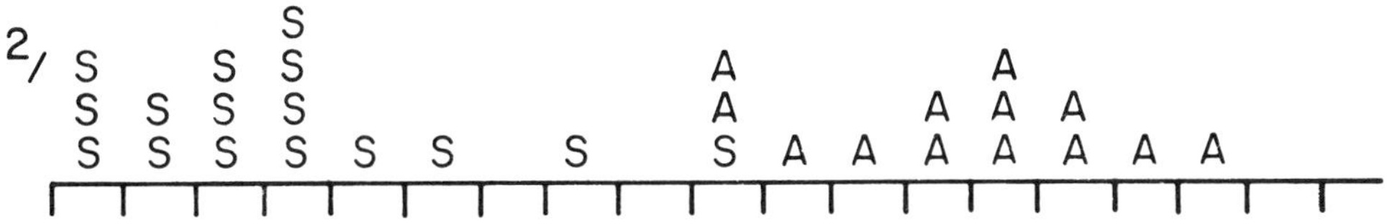

Figure 2. Classification of autosomic recessive subjects, girls (A) and boys (A) and X-linked recessive boys (S) on a discriminant axe, taking into account three disruption indicators (grouped in two variables) on the basis of a discriminant function with two variables and two groups. The units are logarithmic units.

kept constant in both groups. It is, therefore, not possible to explain the difference in the performance of the two groups on the grounds, for example, of the extent of physical impairment; this does not change from one group to another. The difference between the two groups must, therefore, be supposed to be an effect of genetic difference.

2. The question arises as to whether these two groups are representative of patients with DDB I type of muscular dystrophy. This is nonsense as these subjects were carefully selected. The extension of the results presented to the entire population with muscular dystrophy is limited by the effects of interaction which appear between the independent variables (the genotype) and the variables which have been kept constant here (age, mental deficiency, extent of physical impairment).

3. It must also be noted that within the groups characterized by their reactions to parasite stimulation, individual differences were observed.

In conclusion, two points must be noted. This approach allows the carrying out of an intensive study of behavior which is more practical for the psychologist, whereas other methods based on gentic biometrics are necessarily extensive; and the method is necessarily pathological, that is, the categories of the independent variable are, in most cases, made up of alleles which are unfavorable in all situations. It must not be deduced from this that there might be an approach based on favorable genes, nor should it be thought that if one mutation controls mental deficiency, another may control genius.

## REFERENCES

1. Brady, R.O: Inherited metabolic diseases of the nervous system. Science 193: 733, 1976.

2. Carlier, M., Roubertoux, P., Dumont, E., and Perruchet, P.: Analyse génentique des conduites d'acquisition chez des enfants atteints de maladie génique (myopathie de Duchenne de Boulogne): premiers resultats. Communication aux Journées Annuelles de la Société Francaise de Psychologie, Nice, 1977.

3. Demos, J.: Problèmes de l'existence et de la signification des troubles du fonctionnement cérébral dans la myopathie à forme Duchenne de Boulogne. Revue de Neuropsychiatrie de l'Enfant 18: 4-5, 210, 1970.

4. Demos, J., *et al*.: Progressive muscular dystrophy. J Neurol Science 30: 41, 1976.

5. Dufriche, J.: *Thèse de médicine*, Paris, 1976.

6. Feingold, J., Arthuis, M., and Celers, J.: Génétique de l'amyotrophie spinale infantile: existence de deux formes autosomiques récessives. Ann Génét 20 (1): 19, 1977.

7. Frézal, J., Feingold, J., and Tuchmann Duplessis, H.: *Génétique*. Paris, Flamarion, 1971.

8. Gaull, G.E.: *Biology of Brain Dysfunction*. Plenum, New York, 1973.

9. Hamonet, J.: La myopathie en Algerie, Apport de la rééducation fonctionnelle. IVè Journées

Thérapeutiques sur la Myopathie Humaine, Corbeil, 1977.

10. Lamy, M., Royer, P., Frezal, J., and Rey, J.: Maladies Héréditaires du Métabolisme chez l'Enfant. Masson, Paris, 1968.

11. McKusick, V.A.: Mendelian Inheritance in Man. Baltimore, Johns Hopkins, University Press, 1975.

12. Meili, R., and Roubertoux, P.: La structure de la Personnalité. In Fraisse, P. and Piaget, J. (Eds): Traité de Psychologie Expérimentale, Vol. 5, 184, Paris, P.U.F., 1975.

13. Nyhan, W.L.: Heritable Disorders of Amino Acid Metabolism. John Wiley & Sons, New York, 1974.

14. Roubertoux, P. and Carlier, M.: Génétique et Comportements. Masson Publishing Company, Paris, 1976.

15. Roubertoux, P. and Carlier, M: L'apport de la génétique à la psychiatrie de l'enfant et de l'adolescent. In Lebovici, S., Diatkine, R. and Soulé, J. (Eds): Traité de Psychiatrie de l'Enfant, Paris, P.U.F., in press.

16. Slater, E., and Cowie, V.: The Genetics of Mental Disorders. Oxford, Pergamon, 1971.

17. Zellweger, H., and Niedermeyer, E.: Central nervous system manifestations in childhood muscular dystrophy.
Ann Paediatr 205: 25, 1965.

# *Adverse Effects and Social Change*

Raymond Boudon

Sociology, in a way, attempts to establish a permanent utopia by finding the key to social change and thus reducing the uncertainty of the future, whence the various efforts in the 19th century to determine the alleged laws of history. Popper definitely did justice to these attempts when he pointed to the contradictory character of this notion (14). Another obsession of sociology has been to search out, for lack of the undiscoverable laws of history, the fundamental factor of change. Both the production of property and goods and the unequal distribution of wealth and power have been proposed for this role with varying success. Today the majority of sociologists are convinced that the search for laws of history and for the dominant factors in change leads to an impasse.

The only effective research in this direction which is generally accepted today comprises the identification of typical processes of change, with the hope that in time such illustrations can be integrated into the framework of a general theory. Tarde (19) saw in the biological processes of contagion a general analogous model to social change. At present we know that certain social changes apply to this type of structure. Some authors have seen the prime motive power of change in technical and scientific innovation. Thanks to the work of White (23) we are better able to measure the succession of induced social changes through the invention of the metal ploughshare. A classical study by Mendras (10) showed change in terms of a chain of events induced by the introduction of hybrid corn. However, this type of chain, radiating from a source of innovation, is an execptional case in spite of its frequency. Numerous authors have attributed the principal cause of change to conflicts of interest. Some, such as Marx, have considered that the most significant conflicts, that is those with the highest potential for change, or in what the German philosophers call Geschichtlichkeit, meaning "in historicity," concern property and its means of production. Today we tend to place more importance on conflicts related to distribution of power or authority. The work of Touraine (21) and particularly that of Dahrendorf (3) illustrated this tendency in a meaningful manner. Actually, conflicts in the distribution of power are not, even in the modern industrial societies, the only ones with potential for social change. Numerous conflicts, for example, are related to the demarcation of the borderline between public and private matters (see, for example, privatization of abortion and, in general, decline of public regulations regarding sexual, moral, and cultural matters). It is thus impossible to attribute social change to one particular class of conflicts. However, one can go further: although social change may often be associated with a conflict of interest, it does not necessarily have to be due to that. The invention of the wheel was a source of change which was as important as the use of private capital to invest in production. In the same way, change is often the result of a shift in societal and institutional customs and consequently, there may develop a conflict between institutions and social groups rather than between opposing factions within the social groups themselves. In this case the conflict indicates not confrontation but incompatibility.

To summarize, it may be stated that social change is due to conflicts of interest, to shifts or conflicts (the term conflict signifying contradiction in this case) between institutions, and to localized changes in connection with effects of radiation from the source of innovation. The plurality of the processes of change, in other words, appears to be irreducible.

In the following discourse, I should like to isolate an essential mechanism which would represent an important factor in the analysis of social change. This mechanism is that of adverse effects. An adverse effect may exist between two or more individuals, who are searching for a given objective, which gives rise to unlooked for and undesirable effects from the viewpoint of either one or both of the parties.

This idea may be illustrated by a classical example borrowed from the political work of Jean-Jacques Rousseau. Let us, for example, reread the account of the hunting party, which plays an important role in the second _Discourse_ and which certainly provides a key to contradictions (in the logical meaning of the boundary between _Discourse_ and _Contract_ of which Rousseau has been accused but has always vehemently denied). Two savage, rational, egotistic, hedonistic men decided to change their bill of fare, generally composed of hare, and to start hunting for deer (15). Although we live in an affluent society, we could assume that a savage could not by himself overcome a deer. The two hunters, having agreed upon their objective and being sufficiently rational to understand the means to attain this objective, proceeded to lie in wait. However, since the feeling of loyalty, according to Rousseau, is not supposed to have existed in savages, the hunting party failed. In effect, the first of the two savages who saw a hare pass by abandoned the watch. Thus, in spite of the clarity of the objective which they had established, despite their rationality, even though nature did not restrain her production at their disposal, and in spite of the absence of hostility between them, the two savages were incapable of attaining their objective. The analysis reveals an adverse effect which is based upon didactic purity. The corollary of Rousseau's theorem is obvious, even under these optimal conditions; cooperation assumes loyalty. In other words, it is necessary for each of the two friends to agree not to defect. More precisely, each party must agree to accept sanction in case of temptation to defect. Clearly, a social contract is based on this theorem of *Discourse*. Far from the fact that the first of these articles might be a hymn to nature and the second a eulogy of the state of society, the second is a logical sequence of the first. The famous expression "to force man to be free," is a contradiction, not only from the logical viewpoint, but in a dialectic sense: the structure of interaction which presupposes the axiom of *Discourse on the origin of inequality*, results in the fact that the two savages were unable to obtain their objective which they had freely decided to pursue unless they had accepted the force necessary to attain it. The elimination of adverse effects produced by the state of nature is obtained by the introduction of public constraint in a case where it is not realistic to expect a display of loyalty (such as when it concerns not two but a large number of actors).

It is needless to emphasize the relationship between these essential developments of Rousseau's political theory and that which we call the theory of games today. One may attempt, however, to formalize the demonstration of *Discourse* in the following manner. Let us imagine that for each of the two actors a hare has a value equal to two, and one-half a deer has a value equal to three. Each of the two actors is inclined toward two strategies: one a strategy of cooperation and the other, of defection. The matrix of recompense then has the following structure: if the two actors cooperate, each one obtains one-half of the deer. If the second defects while the first cooperates, the first obtains no recompense and the second, a recompense of two. If the strategies are reversed, the recompense is likewise reversed. In the case where both defect, both obtain a recompense equal

| | | Actor 2 | |
|---|---|---|---|
| | | cooperation | defection |
| Actor 1 | cooperation | 3.3 | 0.2 |
| | defection | 2.0 | 2.2 |

to two. One could then imagine that the state of nature is, as Rousseau had assumed, such as to exclude loyalty, so that each of the two actors is uncertain of the other's behavior. This could be translated by supposing that each of them deems it equally probable that the other would adopt either of the two strategies. Under these conditions, actor one hopes to gain (3 x 0.5) + (0 x 0.5) = 1, by choosing the cooperative strategy and of (2 x 0.5) + (2 x 0.5) = 2 by choosing defection. Naturally, this applies to the second actor as well, so that each one has an interest in defection. However, if such be the case, each would receive a lower recompense than he would have received had he been cooperative. Naturally the paradox would not occur if the values for the hare and one-half of the deer were assumed to be wider apart, such as one and four, for example. The analysis shows, however not that the hunting project would necessarily have failed but that it could fail under certain conditions which are at the same time against and for the will of the actors. The only way for the hunters to avoid this trap is to forbid defection, that is, to insure themselves against the temptation of defection by accepting the course that such prohibition be sanctioned.

In fact, it must be observed that the structure of the interaction illustrated by the hunting party of _Discourse on the origin of inequality_ closely resembles the well-known structure of interaction of the games theory, the dilemma of the prisoner. Now the only way to solve the dilemma for actors falling into this trap is to accept freely the constraint which obliges them to cooperate.

To summarize: Rousseau's analysis demonstrated that a fundamental social change, namely the institutionalization of the constraint, could result in adverse effects caused by a state of nature, that is, by situations where one acted as he pleased.

## AN INTERESTING CASE OF ADVERSE EFFECTS OR HOW EQUALITY CAN PRODUCE INEQUALITY

Rousseau's example is of a theoretical nature. However, it is easy to illustrate the relationship between adverse effects and social change by numerous examples relevant to characteristic concrete processes of industrial societies. I shall expand in some detail on an example borrowed from the sociology of education. This example suggests that the crisis which has characterized the systems of education in industrial societies for the past ten years and which has not been eliminated is due to an accumulation of adverse effects which, aside from its positive effects, have been engendered by the development of education.

The analysis of the consequences of the considerable development of the amount of schooling, which has characterized most countries over a period of several decades, is one which seems to be both difficult and controversial. Everyone knows that sociologists, like politicians, have for some time seen in the development of education the best instrument for a policy of social equality. The economists in the educational field still subscribe to the notion that development in education leads to a reduction in the inequality of wages. Recently, sociologists have seen, by an increase in the amount of education, a means for increasing social mobility as well.

My purpose is not to treat the problem of the consequences of increasing the amount of schooling overall. With respect to the influence of this factor upon the distribution of revenues, I shall be content to refer the reader to the work of Thurow (20). It shows that if it be assumed that the structure of employment is determined, only to a small extent, by the modification of the amount of education, a moderate prolongation of schooling does not lead to a reduction but to an increase in economic inequalities. More precisely, if we confine ourselves to a distinction between the three classical divisions of schooling (primary, secondary, higher), it is shown 1) that the variance of wages conforming with each of these levels tends to decrease; and 2) that the mean wages conforming with the three levels tend to differ more from each other.

Production of evidence of this perverse effect which contradicts the frequently advanced propositions by economists in the field of education results in the hypothesis, seemingly reasonable at first glance, that the structure of employment (more exactly, the structure of wages) is modified more slowly than that which we call the schooling structure, namely the distribution of individuals as a function of their level of schooling (22).

The well founded hypothesis by Thurow is demonstrated _a posteriori_ in the following observation in the United States, in the years between 1949 to 1969:

1) _A reduction in inequalities of schooling_. To measure this, Thurow used the Gini-Pareto procedure: if N be the total number of years of education characterizing a population at a given time and $n_1$, $n_2$,....$n_{10}$ be the total number of years corresponding respectively to 10% of the least schooled population, then 10% of the next better schooled, and finally the 10% receiving the highest schooling; then it emerged that part of the total amount of education representing the 10% of the least schooled increased between 1949 and 1969, whereas the part of the 10% representing the highest education decreased.

2) _An increase in economic inequalities_. In effect, between 1949 and 1969, the proportion of the total wages paid to the 10% which were least well educated showed a tendency to decrease, whereas the part of the 10% paid to the best educated increased. Thus, the development of the amount of schooling was accompanied at the same time by a reduction of schooling inequalities and of an increase in economic inequalities.[1]

---

[1] Thurow reveals that there is not only a correlation between these phenomena, but a relationship between cause and effect. We refer you to his text (20) for this demonstration.

From another viewpoint, American statistics reveal that in addition to conforming with consequences resulting from Thurow's hypothesis, between 1949 and 1969 there was evidence of 1) a reduction in the variance of wages of each of the three levels of schooling; and 2) a divergence of means in the wages corresponding to the three levels.

Let us now consider the question of the influence of expansion of the amount of schooling upon social mobility. Thurow concluded that the development of an educational system is not necessarily accompanied by attenuation of economic inequalities, but is quite the opposite; in an analogous fashion it can be shown that there is no reason for an increase in mobility, even if an attenuation of chances for schooling be assumed.

The underlined phrase calls for clarification. It is sometimes admitted in the so-called critical sociology that the school system affects not the attentuation but the reinforcement of inequalities due to birth and that, in consequence, the expansion in the amount of schooling does not have to be associated with a democratization of education. True, it is possible to produce certain statistics, relevant to certain countries and preferably concerned with very brief periods, which show that an inequality of chances for an education does not manifest a tendency toward attenuation. However, if periods of 20 years, 10 years, or even less, be considered, and if the overall development of inequalities in schooling in industrial societies be examined, a general attenuation of educational inequalities will be observed. On this point the demonstration by Thurow in the United States could be criticized, inasmuch as it does not take into account the evolution of the demographic structure of the American population between the periods considered. However, the impressive statistical documentation assembled by the O.C.D.E. regarding the problem leaves no doubt: in industrial societies, and particularly in liberal industrial societies, schooling inequalities manifest a constant tendency toward attenuation (12).

The problem is thus to find out why the expansion of the amount of schooling on the one hand, and the attenuation of inequality of chances for an education on the other hand, are not associated with an attenuation in chances for social inequality, or, if more scholarly language be preferred, an increase in social mobility.

To answer this question, I have developed in the equality of chances (1) a theory which is impossible to expound in detail here. I shall merely present the model to which it leads. Grosso modo, the model consists of three logically linked parts, or, if preferred, of three moments:

1) In the first moment it is assumed that the distribution of individuals of a population in relation to scholastic success varies as a function of the social class of origin. It should be conceded, moreover, that each social class is associated with a field of decision which determines the probability that for an individual belonging to a certain social class and manifesting a given level of scholastic success, a given aptitude or inaptitude for learning, etc., selects this or that course in education (examples: entrance into long-term secondary schooling as opposed to short-term higher education, continuation of studies as opposed to entrance into an active way of living, etc.). It is also assumed that the steps in the course of education and, more generally, the structure of the points of bifurcation marked out by the course could vary from one school system to another and could develop in time.

Let us state that the formalization of the propositions which have just been presented briefly lead to a model by which it is possible to reconstitute the future education from a hypothetical cohort. Thus, let us consider three social classes (superior, average, and inferior); let us imagine that one cohort of 100,000 students end their elementary schooling at a given time and let us assume that, among these students, 10,000 are in the superior class, 30,000 are in the average class, and 60,000 are in the inferior class. A convenient formalization of the preceding propositions and an appropriate choice of parameters permit the determination of the number of individuals who, in each social class, will attain each one of the levels determined by the school system considered.

Table 1 represents the result of the application of the model in a particularly simple case: a total of six school levels were distinguished. Columns 1, 3 and 5 show the proportions of individuals who, in each social class, attain each of the levels of education. The three other columns show the accumulated proportions from the bottom upwards.

It will be seen that the results obtained by the model are structurally consistent with the data provided by school statistics; the inequalities with respect to secondary education, distinguishing the three social classes, are considerable; the inequalities with respect to higher education are still more marked.

2) The second moment in the construction of the model permits the transition from statistics to dynamics. We shall assume a change of characteristics of the fields of decision over the time interval. In the most simple case it is conceded that the probabilities of survival in the educational route leading to higher education increase. It is further assumed that this increase becomes greater as the probability where it applies becomes smaller. Thus, let us imagine that for a given level of educational success, at a given age, the probability of survival at a given point of bifurcation, at a certain time, is equal to p. It might be assumed that at the next time interval, it is equal to p + (1-p)a, where <u>a</u> is a positive coefficient less than 1. In Tables 2a, 2b, and 2c, the principal results obtained by the model are presented, when the hypotheses and parameters leading to Table 1 are used and where, in addition, the dynamic hypothesis which has just been enunciated is introduced. The three parts of Table 2 correspond to the distributions obtained at three successive periods. Thus, let us say that Table 1 corresponds to period $t_0$ and Tables 2a, 2b, and 2c[2] correspond respectively to periods $t_1$, $t_2$ and $t_3$.

TABLE 1

Level of studies as function of social class

| educational level | social class | | | | | |
|---|---|---|---|---|---|---|
| | $C_1$ (superior) | | $C_2$ (average) | | $C_3$ (inferior) | |
| | (1) | (2) | (3) | (4) | (5) | (6) |
| 1. end of higher education | 0.1967 | 1.0000 | 0.0340 | 0.9999 | 0.0053 | 1.0000 |
| 2. higher education | 0.0905 | 0.8033 | 0.0397 | 0.9659 | 0.0104 | 0.9947 |
| 3. end of secondary education | 0.0618 | 0.7128 | 0.0357 | 0.9262 | 0.0118 | 0.9843 |
| 4. second cycle of secondary educ. | 0.1735 | 0.6510 | 0.1396 | 0.8905 | 0.0653 | 0.9725 |
| 5. first cycle of secondary educ. | 0.2775 | 0.4775 | 0.3609 | 0.7509 | 0.3072 | 0.9072 |
| 6. primary education | 0.2000 | 0.2000 | 0.3900 | 0.3900 | 0.6000 | 0.6000 |
| Total | 1.0000 | | 0.9999 | | 1.0000 | |

TABLE 2
Levels of studies as functions of social class at three successive periods $t_1$, $t_2$ and $t_3$

| educational level | social class $C_1$ (superior) | | $C_2$ (average) | | $C_3$ (inferior) | |
|---|---|---|---|---|---|---|
| | (1) | (2) | (3) | (4) | (5) | (6) |
| a) $t = t_1$ | | | | | | |
| 1. end of higher education | 0.2319 | 1.0001 | 0.0491 | 0.9999 | 0.0092 | 1.0001 |
| 2. higher education | 0.0947 | 0.7682 | 0.0490 | 0.9508 | 0.0153 | 0.9909 |
| 3. end of secondary education | 0.0629 | 0.6735 | 0.0418 | 0.9018 | 0.0164 | 0.9756 |
| 4. second cycle of secondary educ. | 0.1707 | 0.6106 | 0.1526 | 0.8600 | 0.0832 | 0.9592 |
| 5. first cycle of secondary educ. | 0.2599 | 0.4399 | 0.3564 | 0.7074 | 0.3360 | 0.8760 |
| 6. primary education | 0.1800 | 0.1800 | 0.3510 | 0.3510 | 0.5400 | 0.5400 |
| Total | 1.0001 | | 0.9999 | | 1.0001 | |

TABLE 2 (continued)

| educational level | social class $C_1$ (superior) (1) | (2) | $C_2$ (average) (3) | (4) | $C_3$ (inferior) (5) | (6) |
|---|---|---|---|---|---|---|
| b) $t = t_2$ | | | | | | |
| 1. end of higher education | 0.2689 | 1.0002 | 0.0680 | 1.0000 | 0.0151 | 1.0000 |
| 2. higher education | 0.0977 | 0.7313 | 0.0584 | 0.9320 | 0.0215 | 0.9849 |
| 3. end of secondary education | 0.0631 | 0.6336 | 0.0474 | 0.8736 | 0.0217 | 0.0634 |
| 4. second cycle of secondary educ. | 0.1662 | 0.5705 | 0.1629 | 0.8262 | 0.1018 | 0.9417 |
| 5. first cycle of secondary educ. | 0.2423 | 0.4043 | 0.3474 | 0.6633 | 0.3539 | 0.8399 |
| 6. primary education | 0.1620 | 0.1620 | 0.3159 | 0.3159 | 0.4860 | 0.4860 |
| Total | 1.0002 | | 1.0000 | | 1.0000 | |

TABLE 2 (continued)

| educational level | $C_1$ (superior) | | social class $C_2$ (average) | | $C_3$ (inferior) | |
|---|---|---|---|---|---|---|
| | (1) | (2) | (3) | (4) | (5) | (6) |
| c) $t = t_3$ | | | | | | |
| 1. end of higher education | 0.3069 | 1.0001 | 0.0904 | 1.0000 | 0.0233 | 1.0000 |
| 2. higher education | 0.0993 | 0.6932 | 0.0676 | 0.9096 | 0.0288 | 0.9767 |
| 3. end of secondary education | 0.0626 | 0.5939 | 0.0524 | 0.8420 | 0.0277 | 0.9479 |
| 4. second cycle of secondary educ. | 0.1604 | 0.5313 | 0.1703 | 0.7896 | 0.1197 | 0.9202 |
| 5. first cycle of secondary educ. | 0.2250 | 0.3709 | 0.3350 | 0.6193 | 0.3629 | 0.8005 |
| 6. primary education | 0.1459 | 0.1459 | 0.2843 | 0.2843 | 0.4376 | 0.4376 |
| Total | 1.0001 | | 1.0000 | | 1.0000 | |

[2] The model is applicable to an ideal type rather than to a particular industrial society; it is difficult to synchronize the time of the model with the actual time. For a better understanding, it might be assumed that the time interval separating two successive periods of the model is equal to almost five years. A number of developmental curves may then be derived from the model which are close to those observed at the level of scholastic compatibility.

These tables reproduce the structural properties which may be observed at the statistical level of schooling when they are used to obtain diachronic data. It is observed 1) that from one period to another the probability of attaining high levels in the educational system is multiplied by a coefficient which becomes greater as the social origin becomes lower; but 2) that the additional number of individuals who, from one period to the next, attain higher education per one thousand individuals is much lower in the inferior class than in the other classes. These two results agree with the data on scholastic compatibility.

The two first moments of the model will not be stressed further. They are relevant here only in that they permit the determination of 1) the evolution in time with respect to actual figures corresponding to the different educational levels; and 2) evolution in time with respect to the social composition of actual figures corresponding to each educational level.

We shall now approach the third moment of the process which directly concerns the problem of interest: that of the influence of the amount of schooling on social mobility and the reduction of inequalities with respect to education. The two first moments permit the establishment of distributions characterizing the successive cohorts from the view-point of the level of instruction. The third moment defines the mechanism whereby individuals, who attain a given level of instruction, attain a given social status.

Which hypotheses are relevant to this aspect? One hypothesis presents itself, namely that the industrial societies are all, to a certain extent, meritocratic. In other words, it is assumed that, all other things being equal, those who have attained a higher level of education tend to reach a higher social status. It will be seen further how it is possible to explain this proposition in a more precise form.

A second hypothesis states that all other things being equal, those whose social origin is higher tend to reach a higher social level. It may be conceded, according to numerous results of a survey, that individuals possessing the same level of instruction have a better chance to attain a high social level if their social origin is higher. We shall discuss the effect of dominance in this case.

If our objective is to establish a comprehensive theory of mobility, still other factors ought to be introduced. It is well known that individuals, through equal levels of education, aim for the types of careers which are associated with higher social hopes. Thus, young people of average social origin, who have ended their secondary education successfully, aim less frequently toward prestigious careers in medicine and toward the right to have school fellows from a higher social origin.

On the other hand, it is clear that factors of an ecological nature affect the processes of mobility: two individuals whose individual characteristics are similar (same type of social origin, same level of education, same type of schooling, etc.) have all the chances to obtain a different social status according to whether they belong to this or that type of environment. The statistics collected by the O.C.D.E. demonstrate the existence of important regional variations in educational chances, even when the individual characters are controlled.

In what follows we shall consider the most simple case. Let us assume that the geographic mobility is sufficiently important for these ecologic factors to be considered negligible. On the other hand, since the question deals with determination of the extent to which social mobility is affected by an increase in the amount of schooling and attenuation of scholastic inequalities, we may neglect the effect of dominance mentioned above. We shall hence adopt the position of accepting the hypothesis of a purely meritocratic structure. If the development of education and the attenuation of the inequality of scholastic chances were to influence social mobility, it might well be in a society of this type.

The question then becomes one of formalizing a meritocratic hypothesis. For simplification, let us assume that we can distinguish, as formerly, three types of hierarchical social status: $C_1$ (high status), $C_2$ (average status) and $C_3$ (low status). Let us further assume that the social structure is relatively immobile at the time, that is, that the distribution of individuals relative to the three types of status is very nearly constant. In the first and second parts of the model, it was assumed that a cohort attaining the end of elementary schooling was distributed in the following manner: 10,000 $C_1$, 30,000 $C_2$ and 60,000 $C_3$. It will be assumed that this cohort must split up a total of 100,000 social positions, of which 10,000 are at the $C_1$ level, 30,000 at the $C_2$ and 60,000 at the $C_3$ level.

Naturally it is unrealistic to assume that individuals who reach the end of the elementary cycle at the same time would also find themselves competing in the employment market: according to their level of education, they would enter the employment market at different times. Let us say simply that it would be easy to complicate the model in such manner that the individuals who are assumed to be in a competitive situation present themselves in the employment market in a synchronized manner. However, this will not modify the conclusions of the analysis.

To translate the meritocratic hypothesis, it would then be sufficient to consider that the individuals will attain their designated social status according to an unequal process favoring those with a higher level of education. Thus, at time $t_0$, 10,000 positions are available at $C_1$, whereas (cf. Table 1): 0.1967 x 10,000 + 0.0340 x 30,000 + 0.0053 x 60,000 = 3,305 individuals of the original cohort attain a higher scholastic level. It will be assumed that a large proportion of these individuals, 70% for example, attain social positions of a high type ($C_1$). There would then remain 10,000 - (3,305 x 0.70) = 7,686 positions available in $C_1$. It will be assumed that 70% of these positions would go to individuals who have attained the scholastic level immediately lower than the highest level. If we refer to Table 1, it will be seen that we are dealing with individuals who had not finished their higher education. The positions available at $C_1$ will thus continue to be given to candidates with successively lower levels of education.

This being completed, the social positions available at the $C_2$ level (social status at the average level) will then be filled by successively accepting the candidates with the highest scholastic level, then those with the scholastic level immediately below that and continuing thus to the lowest level. Naturally, it must be taken into account that some of these candidates have already been placed at the $C_1$ level. Again it will be assumed that the meritocratic parameter will be equal to 70%.

The mechanism of distribution is, as will be seen, very simple and it is doubtless unnecessary to present it at length. It should be noted, however, that in certain cases it could happen that the number of positions available at a given social level might be less than the number of candidates. Thus, it is easy to see by referring to Table 1 that, at the time $t_0$, the number of individuals who have not gone beyond the level of the first cycle of secondary education, namely (0.2775 x 10,000) + (0.3609 x 30,000) + (0.3072 x 60,000) = 32,035 is higher than the number of positions available at $C_2$ when all the individuals with a higher scholastic level have been selected. Let x represent this number. It will be assumed in this case, for logical reasons which we are unable to develop in the framework of this article, that the meritocratic parameter is applicable to x. The number of positions of type $C_1$ that would be apportioned to individuals who had attained only the first cycle of secondary schooling would then be equal to 0.70.

Application of the procedure which has just been described leads to the results presented in Table 3. This table shows the number of individuals who, having a given level of education, attain each of three types of social status at four different periods. The meritocratic parameter was assumed to be uniformly equal to 70% at four periods. The only element which varied from one period to another is thus the distribution of scholastic levels which characterize each of four cohorts.

The consequences of the general elevation of scholastic levels with respect to the relationship between the scholastic level and acquired status are relatively complex. An examination of Table 3 will show:

1) that the higher scholastic levels ($S_1$: higher education terminated and $S_2$: higher education not terminated) are associated with the structure of chances which remain stable over a period of time;

2) that the structure of chances associated with the $S_3$ scholastic level (termination of secondary schooling) is at first glance constant and just as favorable as that characterizing the two higher levels. Nevertheless, at the last time period, this structure suddenly becomes less favorable. The individuals who do not go beyond this scholastic level find that their chances of attaining the highest social status are considerably reduced, whereas chances of attaining the intermediate social level and the inferior level are accordingly increased;

3) that the structure of chances associated with the inferior levels, namely $S_4$ (second cycle of secondary schooling), $S_5$ (first cycle of secondary schooling and $S_6$ (elementary schooling) deteriorates in a continuous fashion over a period of time. It may, however, be seen that this deterioration becomes more pronounced as the level of schooling becomes relatively higher. Thus, in the first period, the schooling level $S_1$ is associated with a rather significant probability (0.2920) of attaining the superior social level $C_1$. At the fourth period, this probability is less than 1/10th as great (0.0249). In return, the degradation of the structure of chances associated with $S_5$, and particularly $S_6$ levels, is slower.

It would naturally be possible to obtain these results by means of an abstract analysis. We have

TABLE 3
Actual numbers and proportion of individuals attaining each of three social levels as a function of scholastic level at four different periods $t_0$ to $t_3$

| scholastic level | | social status $C_1$ | | $C_2$ | | $C_3$ | | Total |
|---|---|---|---|---|---|---|---|---|
| $t_0$ | $S_1$ | 2,213 | (0.7000) | 694 | (0.2100) | 298 | (0.0900) | 3,305 |
| | $S_2$ | 1,904 | (0.7000) | 571 | (0.2100) | 245 | (0.0900) | 2,720 |
| | $S_3$ | 1,678 | (0.7000) | 503 | (2.2100) | 216 | (0.0900) | 2,397 |
| | $S_4$ | 2,874 | (0.2920) | 4,878 | (0.4956) | 2,090 | (0.2124) | 9,842 |
| | $S_5$ | 862 | (0.0269) | 16,345 | (0.5102) | 14,828 | (0.4629) | 32,035 |
| | $S_6$ | 369 | (0.0074) | 7,009 | (0.1410) | 42,323 | (0.8516) | 49,701 |
| Total | | 10,000 | | 30,000 | | 60,000 | | 100,000 |
| $t_1$ | $S_1$ | 3,041 | (0.7000) | 912 | (0.2100) | 391 | (0.0900) | 4,344 |
| | $S_2$ | 2,334 | (0.7000) | 701 | (0.2100) | 300 | (0.0900) | 3,335 |
| | $S_3$ | 2,007 | (0.7000) | 602 | (0.2100) | 258 | (0.0900) | 2,867 |
| | $S_4$ | 1,833 | (0.1625) | 6,611 | (0.5862) | 2,833 | (0.2512) | 11,277 |
| | $S_5$ | 550 | (0.0164) | 14,822 | (0.4431) | 18,077 | (0.5404) | 33,449 |
| | $S_6$ | 235 | (0.0053) | 6,352 | (0.1420) | 38,141 | (0.8527) | 44,728 |
| Total | | 10,000 | | 30,000 | | 60,000 | | 100,000 |

TABLE 3 (continued)

| scholastic level | | social status $c_1$ | | $c_2$ | | $c_3$ | | Total |
|---|---|---|---|---|---|---|---|---|
| $t_2$ | $s_1$ | 3,944 | (0.7000) | 1,184 | (0.2100) | 507 | (0.0900) | 5,635 |
| | $s_2$ | 2,813 | (0.7000) | 844 | (0.2100) | 362 | (0.0900) | 4,019 |
| | $s_3$ | 2,348 | (0.7000) | 705 | (0.2100) | 302 | (0.0900) | 3,355 |
| | $s_4$ | 627 | (0.0495) | 8,421 | (0.6653) | 3,609 | (0.2851) | 12,657 |
| | $s_5$ | 188 | (0.0055) | 13,192 | (0.3871) | 20,698 | (0.6074) | 34,078 |
| | $s_6$ | 80 | (0.0020) | 5,654 | (0.1405) | 34,522 | (0.8576) | 40,256 |
| Total | | 10,000 | | 30,000 | | 60,000 | | 100,000 |
| $t_3$ | $s_1$ | 5,025 | (0.7000) | 1,438 | (0.2100) | 716 | (0.0900) | 7,179 |
| | $s_2$ | 3,324 | (0.7000) | 998 | (0.2100) | 427 | (0.0900) | 4,749 |
| | $s_3$ | 1,156 | (0.2995) | 1,893 | (0.4904) | 811 | (0.2101) | 3,860 |
| | $s_4$ | 346 | (0.0249) | 9,484 | (0.6825) | 4,065 | (0.2926) | 13,895 |
| | $s_5$ | 104 | (0.0031) | 11,331 | (0.3325) | 22,639 | (0.6644) | 34,074 |
| | $s_6$ | 45 | (0.0012) | 4,856 | (0.1340) | 31,342 | (0.8648) | 36,243 |
| Total | | 10,000 | | 30,000 | | 60,000 | | 100,000 |

preferred to use the method of simulation (arithmetic analysis of a model), with the object of presenting the statement in a more concrete fashion (5). From an intuitive point of view the reasons for the phenomena which have just been presented are quite clear: the social structure (distribution of available social status) has formerly been considered stable; in addition, the scholastic structure (distribution of individuals as a function of scholastic level) shows a tendency to skew toward a rise, the increase in actual figures from one period to another being greater when the higher levels of schooling are considered. It follows that the available positions at the higher social level are distributed, with rapidly rising frequency, among individuals whose scholastic levels are highest. After a certain period of time this is accompanied by a sudden degradation in the structure of chances associated with the intermediate scholastic levels, a degradation which is gradually reflected at the inferior levels.

The consequences of the model, from the viewpoint of social mobility, now remain to be examined. Table 2 shows, at each period, the proportion of individuals who, with a given social origin, attain a given scholastic level. Table 3, in turn, shows the proportion of individuals who, with a given scholastic level, attain one of three types of social status. Since the hypothesis was postulated that the status of the destination was determined exclusively by merit and that the effects of dominance were negligible, in order to obtain the matrices of intergeneration mobility corresponding to each of four periods, it would be appropriate to multiply the matrices in Table 2 by those in Table 3, corresponding to each of the four periods. The results are presented in Table 4.

The following remarks may be deduced from this table:

1) In the first place it may be seen, as we have stated, that the structure of mobility evolves very little from $t_0$, the first period considered, to $t_3$, the last period. The probabilities contained in each of the four tables are practically

TABLE 4
Tables of mobility engendered by the model at four periods $t_0$ to $t_3$

| social origin category | | social category of destination $C_1$ | $C_2$ | $C_3$ | Total |
|---|---|---|---|---|---|
| $t_0$ | $C_1$ | 0.3039 | 0.3290 | 0.3670 | 0.9999 |
| | $C_2$ | 0.1299 | 0.3313 | 0.5387 | 0.9999 |
| | $C_3$ | 0.0510 | 0.2795 | 0.6697 | 1.0002 |
| $t_1$ | $C_1$ | 0.3056 | 0.3226 | 0.3719 | 1.0001 |
| | $C_2$ | 0.1304 | 0.3266 | 0.5428 | 0.9998 |
| | $C_3$ | 0.0505 | 0.2829 | 0.6666 | 1.0000 |
| $t_2$ | $C_1$ | 0.3107 | 0.3174 | 0.3722 | 1.0003 |
| | $C_2$ | 0.1323 | 0.3237 | 0.5440 | 1.0000 |
| | $C_3$ | 0.0488 | 0.2852 | 0.6660 | 1.0000 |
| $t_3$ | $C_1$ | 0.3080 | 0.3198 | 0.3723 | 1.0001 |
| | $C_2$ | 0.1319 | 0.3246 | 0.5435 | 1.0000 |
| | $C_3$ | 0.0494 | 0.2855 | 0.6650 | 0.9999 |

identical from one period to another. This result, which can be justified mathematically, appears to be paradoxical. Let us recall that the results in Table 4 are the consequence of a model that assumes the following, between $t_0$ and $t_3$: a) a marked growth of the actual scholastic figures at the highest degrees of the hierarchy of scholastic levels; b) a non-negligible attenuation of the inequality of chances for an education; c) a marked modification over time of the scholastic structure contrasting with a fixed social structure.

Intuitively, it is tempting to conclude that these different factors lead to changes in the structure of the mobility. The analysis shows, however, that this is not so. The model engenders simultaneously marked changes in the scholastic structure and a non-negligible attenuation in the inequality of chances for education; but these changes have no effect on modifying the mobility structure.

2) Let us now consider the slight changes in the mobility structure which occur between $t_0$ and $t_3$. Between $t_0$ and $t_2$, a slight tendency toward an increase in autorecruitment is observed at the level of the $C_1$ class. At the same time, the decreasing mobility between $C_1$ (the superior class) and $C_3$ (the inferior class) also rises slightly. On the other hand, between $t_2$ and $t_3$, the autorecruitment characteristics in mobility structure are at the same time low in amplitude and of erratic appearance.

Why should this be so? Without going into the mathematical analysis of this phenomenon, we might attempt to approach this from an intuitive viewpoint. Let us, for example, consider the individuals of a high social level and examine the effects upon their social chances of structural changes postulated by the model between $t_0$ and $t_3$. Between these two extreme periods the individuals in this category attain, on an average, the highest scholastic levels. Thus, at $t_0$, out of 10,000 young people of high social origin 1,967 attained the highest scholastic level (completed higher education); at $t_3$, the figure is 2,689 (Tables 1 and 2). This obviously means that the number of those who did not attain the highest scholastic level tended to decrease. However, at the same time, due to the effect of a generalized increase in demand for education, the structure of chances associated with inferior scholastic levels, and above all the means, tended to deteriorate over time. Now the number of individuals of high social origin who attained only average scholastic levels remained high between $t_0$ and $t_3$. Thus, a sort of compensatory effect was produced which engendered a stability over time of the characteristic structure of mobility in individuals of the superior social class. It is possible to reproduce the same type of analysis in the case of individuals of an average or inferior social origin. In all cases, the practically complete stability of the mobility structure results from the fact that the generalized increase in the demand for education is accompanied by complex compensatory effects.

3) Let us now return to Table 3 (probabilities of obtaining a $C_1$ social level when a scholastic $C_1$ level is attained). This table shows that the structure of chances which characterize the scholastic levels tend to become differentiated over time. Thus, the structure of chances associated with levels $S_1$ and $S_2$ remains stable; on the other hand, the structure associated with levels $S_5$ and $S_6$ deteriorates. This result confirms Thurow's conclusions: the relationship between social aspirations and scholastic level tends to become more and more marked. This factor is doubtless, in part, at the bottom of the generalized increase of the demand for education which characterizes the industrial societies. However, the compensatory effects prompted by this increase result in the fact that the mobility structure remains practically unchanged.

The principal conclusion of the analysis is that there is no reason whatsoever to expect that this considerable increase in the demand for education in the industrial societies should be related to an increase in social mobility, even though it is accompanied by an indisputable reduction in the inequality of chances due to education. The preceding model shows, on the contrary, that under extremely general conditions the confusions which characterize the scholastic structure are normally associated with a higher stability of the mobility structure. For this conclusion to be invalidated unrealistic propositions must be introduced: to admit, for example, that the reduction in inequalities due to education has been much more rapid than it has been in reality; or to assume that certain changes in the social structure, due particularly to technological development, have been extremely rapid (with the same rhythm as those which characterize the development of scholastic structure).

It must be pointed out, with respect to this last point, that in the presentation of the model we have assumed that the social structure had been fixed at the time. This hypothesis is manifestly excessive. The technological development could very well be associated with modifications in the socioprofessional structure, by reducing the proportion of manual manipulations of a repetitive character. However, it is important to state that the conclusions resulting from the preceding model remain true even if a modification be assumed in the social structure over time. Thus a hypothesis may be put forward that, from one period to another, the number of positions available at the superior ($C_1$) social level increases, whereas the number of positions at the inferior ($C_2$) decreases; but unless it is assume that changes in the social structure are as rapid as changes in the scholastic structure, it will be concluded that the mobility structure must remain practically stable over time.

To summarize: under extremely general conditions, the development of education is not associated with a reduction of this form of inequality, which is particular and essential, of social chances (dependence upon social status of son in relation to father's social status), even when it is accompanied by a reduction in inequality of scholastic chances. Perhaps this result contributes to the explanation of the conclusions, also unexpected, of the famous work by Lipset and

Bendix on social mobility: when these authors undertook, toward the end of the 1950's, to compare the mobility in different industrial societies and drew conclusions as to the similarity of the amount of mobility in nations who nevertheless differed markedly in other respects, whose systems of stratification were different, and where education was very unequally developed (9). More than ten years later, in an article on "Public Interest," Lipset confirmed this result (8).

To conclude, I should like to borrow from Thurow: "In any case, I would argue that our reliance on education as the ultimate public policy for curing all problems, economic and social, is unwarranted at best and in all probability ineffective" (20)--ineffective, if one is to accept Thurow's analysis, from the viewpoint of the reduction of economic inequalities, the expansion of education being probably no more effective from the viewpoint of the reduction of social immobility. In this respect, the principal effect of the increase in the demand for education appears to be to require from the individual an increase in schooling without an end in counterpart to social aspirations which remain unchanged.

Perhaps--this is the central hypothesis toward which this analysis leads--these unexpected and perverse effects, may be the underlying cause of the crisis in the educational systems of the industrial societies because unexpectedly they have provoked an immense disenchantment with the social and political virtues of education or because perversely they have provoked a feeling of doubt regarding the finalities of systems of education and inability regarding their management.

## ANOTHER CASE: HOW ONE CAN DISSOCIATE ONESELF FROM HIS OWN INTERESTS?

My second example deals with what may be called the logic of participation. One of the sources of social conflicts is consequently social change inherent in the opposition of interests between groups, but perhaps even more important is the perverse effect which under very general conditions leads, on the contrary, to a passive acceptance by the members of one group of a situation which is against their own interests.

The study of this perverse effect is the object of a book by Mancur Olson, The Logic of Collective Action (13). The impertinent proposition, which Olson seeks to demonstrate in this book, is in effect that a group of unorganized individuals with a common interest, conscious of this interest and having the means to realize it, under general conditions, did nothing to promote it. The interest in common, even when this interest was evident to all, did not suffice to provoke a common action which would promote this interest for all. The logic of collective action and the logic of individual action do not add up to one but to two.

Such a proposition would have surprised neither Rousseau, Marc, Michels, nor a large number of political and sociologic classical philosophers. It would not have shocked any economist. Olson's theory is, in fact, an ingenious application of well-known results in economy to phenomena generally considered to be in the department of sociology or political science.

To introduce Olson's demonstration, I shall repeat one of the examples, simplified here, which he used in the application of a method of analysis.

The question is to discover whether an unorganized group, being conscious of its interest (whose interest is not compatible with that of any other group), disposing of the resources necessary for its promotion would act in a manner necessary to promote its interest as would an individual under these circumstances.

Let us point out at the outset that the language used in the preceding paragraph is incorrect from the moment that the metaphysics of group--individual identification is rejected: it is then not permissible to talk of a "group being conscious of its interest" nor employ any expression which implicitly designates the group as an independent entity of individuals who comprise it and which is transcendent in relation to them. Restated correctly, Olson's question is as follows: let us assume that an imaginary group of individuals with a common interest, who are conscious of this interest and each of whom could contribute to the realization of this interest: will they act effectively in keeping with this common interest?

In order to establish the ideas, let us take a simple example which has the advantage of not being emotionally charged. This example is one of those used by Olson himself. Let us imagine that a group of landowners are interested in obtaining a reduction in their real estate taxes. The group is composed of 10 members (N = 10). Each owns property valued at 10 francs. Each landowner must pay a tax of four francs. Let us then imagine that if they wish to launch a campaign for their own benefit or to use pressure of some sort on the fiscal authorities, they could obtain a tax reduction. To make the example more concrete, let us assume that if each participated in a collective action each would obtain a reduction of 50% in his tax assessment and that the reduction obtained would be a function of the number of participants in the collective action in such fashion that with _n_ participants the tax reduction would be equal to 5n%. This means that if nine members participated in the collective action, each of the 10 members would obtain a reduction of 45% of the initial tax, of four francs; similarly, if the number of participants be equal to 8, 7.6...,1, 0, the reduction for each of the 10 members would be 40, 35, 30,..., 5, and 0%. Finally, let us suppose that participation in the collective action involved a cost (loss of time, financial participation, etc.), which is measurable and which may be fixed at one franc per individual.

Thus, if all the members participated in the collective action they would obtain a total benefit valued at 20 francs (since each one of the 10 members would reduce his fiscal tax from four to two francs) which would cost them a total of 10 francs (one franc per individual).

If the group could be compared with one individual, it is clear that the group-individual would have a gain amounting to 20 francs by paying 10 francs. However, a group, even if it is composed of individuals with a common interest, is not an individual. Let us actually look into the reasoning of any one of the group members: Ego. Ego has two possible options: to pay his portion, that is to contribute to the collective action or not to contribute. Naturally, the advantage that he would gain from one or the other strategy would depend on the behavior of the others. Let us first postulate the hypothesis according to which nine members of the group, other than Ego, contribute to the collective action. In this case, if Ego himself does not contribute, the tax reduction would be 45%, that is, that Ego will have to pay a tax of 4 - (0.45 x 4) = 2.20 francs. By not contributing to the collective action, he thus gains 4 F - 2.20 F = 1.80 F. If he does contribute, the tax reduction would be greater for all: with 10 tax payers the tax is reduced by 50%. The tax then drops from four to two francs and each gains two francs. Like the others, Ego then gains two francs in the form of a tax reduction; but his participation in the collective action costing him one franc, his net benefit in the case where he, like the nine others, participates in the collective action amounts to one franc. It is, therefore, clear that, in the hypothesis where nine others participate in the collective action, Ego himself has an interest in nonparticipation.

Let us now assume that there are eight participants in the collective action, other than Ego. If Ego does not participate in the action, he will benefit from the 40% tax reduction obtained by the eight others. He would then have a net gain of 4 x 0.04 = 1.60F. If he participates in the action, the reduction for each would be 45%, namely 1.80 francs, but Ego will then have to deduct the cost of his participation in the action; his net gain in this case would be 0.80 francs.

By continuing to reason in the same manner, it will be seen that, no matter what the number of participating members is, Ego will have an interest in nonparticipation; but since it is assumed that

Ego is no different than any of the other members of the group, it is clear that if there is no advantage for him to pay, this goes for all the others. Thus, each has an interest in nonparticpation in the collective action no matter how many others participate. It follows that no one will participate in the collective action in which each would have an interest that all participate. Owing to the difference of the action of one person, the group will pass up the chance of earning a total benefit of 20 francs in return for 10 francs.

It is evident that this disturbing result is not a product of the arithmetical pecularities of the example chosen; the arithmetic here serves no other purpose than to facilitate the reasoning. In fact, the paradox originates from the fact that the collective action has the effect of producing a benefit (reduction of tax) which is assumed, by its nature, to benefit all the members of the group.

One objection that could be addressed to Olson is that the postulate that the members of the group would behave in a rational manner is an unrealistic viewpoint. The response to this can be that the strictly rational calculation used in the preceding example in no way implies the hypothesis according to which, in reality, the individuals would have an accurate view of the costs and advantages involved by their possible participation in a collective action and would be capable of analyzing the possible benefits which would accrue from one or another line of action. More subtle and more realistic hypotheses would lead to the same global result but would needlessly complicate the demonstration.

In fact, the demonstration assumes little more than adherence to the principle according to which, in many cases but not in all, there is an advantage in trying to explain the behavior of individuals in a society by assuming that they essentially seek to better serve their own interests. As Olson himself has stressed, this postulate is frequently inadequate. However, it is also indispensible for the explanation of a number of social phenomena. That is why it is found to be explicitly and systematically used by numerous highly respected sociologists. However, it is superfluous to attempt to confer upon Olson's theory the legitimacy which springs from authority. His numerous applications are sufficient for his defense. The theory explains, for example, that all taxes are always coercive. Even though it were in the citizens' interest to dispose of the public benefits produced by the taxes, no one would contribute to the production of these benefits if he were not forced to do so, even in the case where there had been unanimity in the use made of these taxes by the public authority. A possible explanation which may also be deduced from the theory, is the relatively painless character of the indirect taxation individually incorporated into the price of merchandise sold to the individual consumer: when paid for the purchase of individual goods or services, its function in financing the collective gains becomes invisible, in contrast to the direct tax which is not associated with individual gains. This distinction between the two forms of taxes evidently has adverse consequences from the viewpoint of social justice, given the often regressive character of taxes on consumption.

Another paradox which may be easily explained in the framework of Olson's theory and with which traditional political sociology has often clashed is the following: how does one explain that political parties never recruit a number of militants who represent more than a modest proportion of the electors who vote for them in an election? Olson's response is simple: a political party may essentially be considered as a furnisher of collective benefits. Consequently, if it be conceded that to enroll in a party and to militate in its favor represents costs, for example, in time or money, Olson's theory applies if each member of the latent group, interested in the production of collective benefits, offers to pay for the others such expenses in the party which interests him, for as long as the party is incapable of exercising coercion on its potential clientele.

How then can we explain the fact that the party can induce certain individuals to enroll and possibly militate in its favor, even if these individuals generally represent a small minority in proportion to the electorate of the party? Olson's response is that generally one party furnishes not only the collective benefits but also the individual benefits. Thus, a party is capable of offering to its members positions of political responsibility within the party and, with decreasing uncertainty, the implantation of elective positions in local or national political life. Naturally, the number of positions which a party can offer is always very small in proportion to the number of electors. These are of the sort that the candidates, members of the party, even when they outnumber the available positions (exactly like the candidates to the École Polytechnique outnumber the candidates admitted) should normally be many fewer than the electors. The latter are content to profit by the collective benefits furnished by the party but have no interest in paying the price.

A parallel provision of individual benefits represents the first possible loophole to Olson's paradox. It explains, for example, the power of certain professional organizations, such as the American Medical Association, about which Olson's analyses are particularly brilliant. This power, according to Olson, is due to the quality of individual benefits furnished by the association to its members, legal assistance in particular, in the case of a professional error. The theory of selective incentives, according to which a good strategy for an organization in the act of expansion consists of associating individual benefits with those which it furnishes for the collective benefit, equally permits the explanation of certain obscure results in matters of union affiliation. How, for example, does one explain that, in spite of the obvious benefits which unions obtain for individuals whose interests they serve, the amount of unionization in such different countries as France and the United States does scarcely exceed 20% on the average? Olson's response is again that so long as the organization furnishes benefits, such as wage increases, which benefit everyone as soon as they are produced, no individual is interested in paying the corresponding price for acquisition of these benefits. This proposition explains that the union membership may be low, even in a case where such action is doubtless effective. From another viewpoint, the global amount of unionization is not zero, and the average of 20% which is seen in France and in the United States is accompanied by a very wide dispersion, certain branches having organizations close to the maximum, others close to the minimum.[3] To explain these different phenomena, it suffices to observe the nature of the individual benefits, according to Olson, which in each case the unions are in a position to furnish, aside from the collective benefits whose provision is the reason for their formation. If these individual benefits would correspond to membership positions of responsibility only, the amount of unionization would generally be modified. If, as in the case of legal assistance for professional errors, the union is able to provide an individual benefit of considerable value which each member might need, it would generally be found that the number of members would increase. These reasons apply, for example, to the insignificant union organization among French students. In this case, the union furnishes collective benefits. In the case where a claim for the increase in grants is effective, it is due to the fact that it has slightly increased for each student the probability of obtaining a grant. On the other hand, one would expect to see a rise in membership where the organization would furnish, for example, apart from the action concerned with wage increases (collective benefits), an assurance of unemployment benefits (individual benefit), or insurance against occupational hazards.

Aside from these possibilities of application and of the new interpretation which it offers for

---

[3] Thus in France the global amount of unionization is 20%, but this amount is equal to 75% of the educational institutions. Cf. "Le Monde."

numerous social and political phenomena, Olson's theory likewise points the way to the reinterpretation of certain classical authors. The lecture which Olson has presented regarding Marx's theory of social change and incidental comments on Durkheim's theory is very stimulating. When Marx conceived the capitalist interests as contradictory with those of the proletariat, he meant to indicate by this that the capitalists and the proletariat constitute two latent groups, each composed of individuals having the same interest, which is contradictory to the common interest of members of the other group. But Olson then objects, the proposition according to which the conflict of interests necessarily leads to the war of the classes is incompatible with coherent utilitarianism, inasmuch as the latent groups, the classes, are fundamentally incapable of undertaking collection action for the promotion of their interests. Existence of classes, yes; war of the classes, no. To this, one can respond that it is not certain that Olson said that Marx had not perceived the problem. The distinction between the class in itself and the class for itself, the infinite hesitations of Marx regarding the notion of class conscience, and the developments in the political organization of classes show that he was at least implicitly conscious of Olson's paradox. Besides, this is not surprising for all political philosophy of the 18th century and all the economy at the end of the 18th century and at the beginning of the 19th, on which Marx had been reared, revolved around the paradoxes of collective action.

However, perhaps the most interesting consequence of Olson's theory from the viewpoint of social change is the proposition to which Robert Michels had given the name of the law of the weight of oligarchy (11). Let us assume, in effect that a large number of nonorganized individuals have an interest in the production of a collective benefit. According to Olson's theory, under very general conditions these indiviuals would be incapable of producing this benefit. If this actually is the case, the persons creating this collective need would invite a contractor for a certain interval who would get the organization started in order to explore the means for satisfaction of this collective need. The provision of parallel individual benefits would permit the contractor to win over his potential public. However, once the organization is created, there is nothing to guarantee that the interests of the individuals which it was supposed to have represented would be effectively served. In effect, if it be assumed that the members of the organization take measures or adopt a policy of which their employers would disapprove if consulted, it will be deduced from Olson's theorem that, unless they are explicitly consulted, they will not manifest their opposition to the policy followed. Such manifestation would correspond to the production of a collective benefit. Now the theorem shows that a nonorganized group is, under general conditions, incapable of producing a collective benefit. Also, the group of employers is disorganized in relation to its organization (if mechanisms of electoral control of the organization be expected, by the "base," who are essentially intermittent and often symbolic).[4] In other words, in the case where the organization which represents them follows a policy which deviates perceptibly from the interests of their employers, they are, under general conditions, incapable of manifesting their opposition. The solution of this latent state of crisis is often actualized by a new contractor who exploits, for his own profit, the market created by the discord and who creates either a new organization oriented toward the same potential public or an opposition to the old organization. In quite a few cases Olson's theorem shows that this leads to an oligarchic management in favor of the employers and according to a better hypothesis, to a rivalry between the competing oligarchies.

Certainly, this law of the weight of oligarchy, which Michels had discovered through an erudite

---

[4] Besides, the form of consultation is generally avoided by employers.

analysis of the history of European political parties in the 19th and at the beginning of the 20th century (11), and which appears to be a corollary (not, however, deduced by Olson himself)[5] of the Olson theory of collective action has, in this context, considerable consequences. In effect, from the moment that the representative agents of group interests show a tendency to take the form of an oligarchy, tension and imbalance result and naturally lead to social change.

As Olson himself has indicated, it is clear that his theory is far from being applicable without restriction to all types of situations, for there are some cases where the costs of collective action are negligible, nonexistent, or negative. Participation in a political manifestation could be a pleasure in itstelf, could break the daily monotony, and could endow the actor with a sense of importance. Of all the situations which lend themselves to a utilitarian analysis, the only ones that derive from Olson's theory are those that effectively imply positive costs. On the other hand, there are evidently some groups that do not derive, except very indirectly or at the price of warping the idea of interest or utility which borders on rendering it empty and tautologic, from a utilitarian analysis. It is understood that charity not only begins at home but often stays at home. It goes without saying that a utilitarian analysis of religious groups, for example, does not go very far. Olson's theory thus finds its application within very precise limits.

## HOW SOCIAL ILLS COULD STEM FROM NO PARTICULAR INDIVIDUAL NOR PROFIT ANY INDIVIDUAL

In the two preceding sections, I have presented in detail two theories depicting how states of crises, of social imbalance, or of tension can

---

5 Michels was cited only once by Olson and that, quite incidentally.

result from the appearance of adverse effects, that is, effects that have not been sought by social agents and do not result from the opposition of interests and of conflicts from which they derive. Once more, I do not claim that these adverse effects are in all cases the cause of social change. There are other types of processes which lead to social change: radiation of a chain of events emanating from a localized source, contradiction[6] between the interests of large groups such as social classes, obsolescence of institutions, wars, conquests, etc.; but the processes of change due to the appearance of adverse effects are of particular interest--in the first place, because they are very frequent, secondly, because they often continue without being perceived, and finally, because they are theoretically more complex than the other types of change. The contradiction between opposing interests is immediately perceptible. It is clear that in a number of social conflicts the gain of one of the partners corresponds to the loss of another. In this case the contradiction between the interests takes the form of a simple logical contradiction (incompatibility of interests) or, as it is called in the language of the theory of games, a game of <u>sum zero</u> (the losses of one correspond exactly to the losses of the other). The adverse effects constitute, by contrast, contradictions of a higher degree of complexity. In Rousseau's hunting party, the two partners are losers. The fact that each family unit manifests a scholastic demand which is superior to that of a comparable family results in the effect of each unit acquiring its social status at a higher cost; also the democratization of scholastic chances has scarcely influenced either the equality of social chances (see social mobility), or the inequality of revenues. The logic of collective action from the viewpoint of the production of collective benefits has the effect of creating structures of oligarchic representations

---

[6] In the sense of : game of <u>sum zero</u>.

which could have negative consequences for all. All of these adverse effects have the structure of games of sum zero. It should be observed that these fundamental structures for analysis of social change are naturally ignored by those who create conflicts against social groups (games of sum zero), the exclusive motive force of change.

In this section, I should like to present, in a more succinct fashion than in the preceding cases, some additional examples which would help in the assessment of the vast area of application of the adverse effects.

## The Desertion and Protestation of Hirschman

Hirschman's small volume, Exit, Voice and Loyalty (6) springs from an observation which in itself is quite common, but is very fruitful, because it contradicts a classical theory. The theory of perfect competition states that when a firm attempts to sell its product above the going price, it results in a flight of consumers toward the competing firms which forces it to return to the initial price. It must then consider the means which permit it to reduce the costs of production, by improving its management, for example, or by increasing its productivity.

Now, in the course of his research on development, Hirschman encountered a case in this context, which was the complete reversal, trait for trait. So long as the Nigerian railways benefited by means of a monopoly, they operated expeditiously. When they were exposed to a route competition, their situation began to deteriorate irreversibly. The explanation of the phenomenon is simple: the beginning of road competition deprived the railway not only of an unfavorable fraction of its users, but of the most exacting of them. Consequently the transition from a situation of monopoly to a competitive situation was accompanied, for the railway company, by not only the appearance of but, as the theory would have it, by the disappearance of an inducement to its modernization.

It may first be concluded by this case that the beneficial action of "the invisible hand" does not necessarily imply a situation of competition. It can also apply to the monopoly situation. This is the central theme of Hirschman's book though the course differs. In this case with respect to competition, it operates by reducing the desertion of the consumer. If this desertion affects the consumers unfavorably, and if the firm alluded to plans to react, the effect is beneficial. If one or the other of these two conditions is not realized, it could result, on the other hand, as with the Nigerian railways, in an unfavorable effect. In the case of the monopoly, the invisible hand acts by inducing not only the desertion but, by an impossible hypothesis, the _protestation_.

## _Desertion and Protestation_ [7]

These are the two fundamental types of reaction to the decline, not only of firms, but equally, as the book subtitle indicates, of organizations as well. In this respect the social sciences, Hirschman observes, are characterized by a singular division of work: the only mechanism taken into account by the economic theory is that of desertion; as for political science, it appears to concede that only protestation deserves its interest. Would it not be more fruitful, Hirschman asks, to draft a theory which, by crossing the border-line between economics and political science, will investigate the conditions of the appearance of desertion and protestation, will analyze their determination as a function of the types of organizations while evaluating their efficacy at the same time? Instead of investigating the effects of desertion on the firms and the protestations on organizations, why not investigate the complementarity of the two mechanisms?

---

[7] "Defection" and "prise de parole" in the French translation.

One of the most brilliant applications of the pair, desertion/protestation, appeared in the critique presented by Hirschman, of the famous model of Hotelling. As is well known, Hotelling (7) had proposed in 1929 the assimilation of a system of political parties with a collection of firms seeking to partition a market among themselves. In the simplest case, the model assumed that the electors were distributed over a linear ideological continuum and that they would vote for the party which adopted the position closer to their own. The figure shown below (where two parties were assumed, A and B) illustrates how simple it is to guess what will happen. Let us, for example, consider the A party: all the electors, ideologically situated between O and M (the center of the segment AB) are closer to A than B. If ideologically they are displaced to the left (toward O) A will lose a certain number of electors: those who will, by this displacement, be closer to B than to A. On the other hand, by their displacement to the right, they can win over a part of the B electors. Similarly, B could not hope to take away his opponent's electors except by displacement to the left, from which it may be concluded that at equilibrium the two parties ought to have positions which are ideologically close.

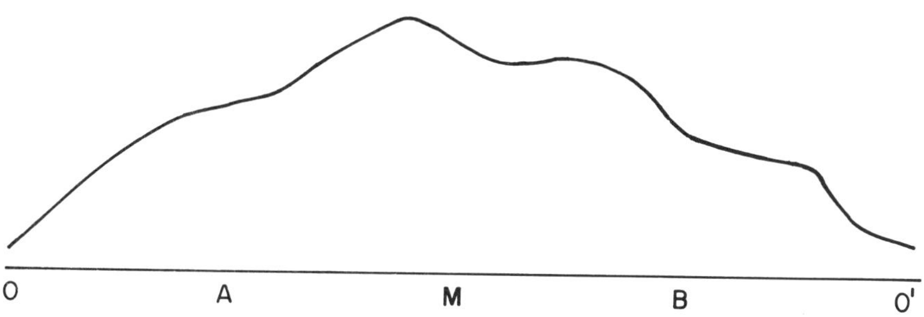


Figure 1. Example of Hotelling model.

It is understandable that Hotelling's model has raised a certain amount of interest. It provided an explanation of the ideological proximity of the two great American parties; it explained that the revolutionary parties could not lay claim to a large clientele unless they exchanged their ideal for a moderate reformism; it explained the frequent but strange phenomenon of the division of the electorate into two parties, which are perceptibly equal in bipolar systems.

The New Deal is associated with a correction by Downs (4) of the Hotelling model. To explain the ideological divergence of the two parties, in this context, Downs introduced the hypothesis that the captive electors (that is, all electors situated to the left of A and to the right of B) could be perceptible of the distance which ideologically separated them from the closest party, and abstain if this distance went beyond a certain threshhold. In economic language, it assumes, contrary to Hotelling, that the elasticity of voting in relation to ideologic distance is not necessarily zero.

Hirschman then asked, "Is it not remarkable that good economists such as Downs and Hotelling exclusively introduce, into their models of mechanisms, desertion? Is it not reasonable to assume that when captive electors see an increase in the ideological distance which separates them from the party closest to them, they would tend to resort to protestation?"

If the hypothesis is accepted, the phenomenon of over-shooting, which characterized the American presidential elections of 1964 and 1968, is better understood. For example, protestation thus explains that the Republican candidate (Goldwater) was selected as too far to the right in 1964, and Democratic candidate (McGovern) was selected as too far to the left in 1968. Another brilliant explanation of the protestation/desertion pair is the interpretation which Hirschman gave of the _law of the weight of oligarchy_ by Michels. In a political system of two parties, he argued, the ideological distance between the parties is on the average higher than in a system of multiple parties. On the other hand, the proportion of captive members ought to be higher in the first case than in the second. In consequence, protestation will play a more important role in relation to desertion in the first case. Conversely, desertion plays a more important role in relation to protestation, as the number of parties increases. Consequently, the multiple party system, because it minimizes the role of protestation, indeed favors, as Michels has shown, the appearance of political oligarchies. But, remarked Hirschman, "It must be observed that in enunciating his famous law of the weight of oligarchy Michels bases it essentially on his experience of the systems of multiple parties in continental Europe."

These remarks and others suggest that the institutions or, as the case may be, the organizations imply by their structure a certain amount of protestation and of desertion. If I am not satisfied with my grocer, it is likely that I should have a tendency to leave him rather than protest, except if it concerns a monopoly situation; if I am not satisfied with the party for whom I generally vote, I would desert it if I could find a competitive party which conforms more with my views; but if I am militant in a party in which I enrolled after mature reflection, it is probable that the desertion would not intervene unless the resources of protestation had been exhausted. The amount of protestation/desertion varies as a function of the cost of enrolling in an organization, the cost of withdrawal, the presence or absence of organizations providing a product or service of substitution, or the facility of access to these competitive organizations and other variables which Hirschman introduced in his list of examples.

The third important concept, which appears in the title of the book, is that of loyalty. It is mainly introduced to account for situations where the cost of leaving an organization is high and where protestation in itself is expensive or ineffective. In this case I should accept without hesitation a certain degree of disagreement in the organization to which I belong; I should remain loyal to it in spite of the distance which separates us. The higher the costs of leaving, the costs of prohibitive protestation, and the lower the efficacy of protestion, the greater r tendency there exists by force of circumstances for loyalty. That is why chauvinism is so widespread even among those who are most critical of the society in which they live; that is why the decision to emigrate (desertion) is always dramatic.

It will be seen that we are far removed from the Nigerian railways, although the figure which inspired the work was encountered verbatim in the industrial societies. Thus the desertion of the comfortable classes toward private schools provoked in the United States an apparently irreversible degradation of the system of public schools; similarly the competition of telegraph and telephone deprived the postal services of the elite of their clientelle and produced an accelerated deterioration of this public service. However, the variety of examples to which Hirschman applied the fundamental concepts of desertion, of protestation, and of loyalty showed the general scope of his theoretical diagram.

Thus, despite the shape of his draft, Hirschman's theoretical diagram enables us to see at a glance, by means of the example of American society, an interesting explanation of the "morosity" of industrial societies. Its indications in this respect are transient and vague. The progressive uniformization of the urban environment of the United States has the effect of reducing the possibilities of desertion; whence the growing intensity of the protestations and eloquence displayed to denounce the universe of the suburbs, doubtless stronger than that which inspired the slums. The Black Power is a substitute for the protestation for the desertion begins at the moment when it becomes clear that desertion by mobility has been an illusion. To this may be added the repetition of a theme which has been discussed previously: the equalization of chances of an education is not accompanied by the anticipated effects from the viewpoint of mobility. The ecological movement developed from the moment that the most sheltered, for lack of power to desert the inconveniences of pollution, had begun to protest. The associations for the defense of consumers encountered a large audience when, because of the general affluence, durable wares took an important place in the family budgets. If desertion is less costly in the case of perishable goods, a protestation becomes evident in the case of durable goods. The political importance of the war against crime became evident when criminality extended to the suburbs, making it impossible for the average classes to desert. Using Hirschman's analysis, the list may be lengthened indefinitely; add to it decentralization and particularisms, especially those of age groups, contracultural phenomena, and a number of other phenomena. All these changes could be analyzed from the dialectic of adverse effects.

## The Costs of Organization According to Buchanan and Tullock

The book by Buchanan and Tullock, The Calculus of Consent (2) takes up the problem posed by Jean-Jacques Rousseau in the Social Contract: what type of representative organization should be adopted to counter the adverse effects of the problem of prisoners? The response of Buchanan and Tullock is first to observe that when the elimination of an adverse effect is dealt with, the costs of this elimination must be considered. As has been shown since Rousseau, these costs could be high. The unsociable man must abandon his natural freedom in order to accede to civil freedom, that is, to accept the submission to constraint and sanctions foreseen in the contract. The matter of interest in the book by Buchanan and Tullock is that he relativizes Rousseau's propositions: in certain cases, the elimination of an adverse effect could be more disagreeable and costly than the adverse effect itself. Thus, below a certain threshold of automobile traffic, it is preferable not to install red lights. Independently of the costs of installation, needless waiting at red lights could, on the whole, be more difficult to tolerate than the inconveniences due to their absence. Naturally, beyond this threshold, the situation is reversed. When the traffic is very heavy, most drivers accept the constraint of red lights with good grace. Although this constraint is really disagreeable, it is less so than the inconveniences which would result from its absence. The difficulty is, of course, that it is impossible to determine the threshold at which one of the solutions must be abandoned in favor of the other. In truth, it is more realistic to substitute, instead of a threshold, a zone of uncertainty. In this zone it is not possible to state with certainty which of the two solutions is preferable to the other.

Analogous considerations are applicable to organizational problems which are more complex than those of traffic lights. When a small number (*n*) of individuals cooperate in the performance of a task, an organization of the democratic type where each one has a fraction, (1/n), of the power of decision could be effective, from the viewpoint that it minimizes the costs of the interdependence. The task is accepted in an adequate fashion and the elimination of divergences of views is not too costly (in time expenditure, for example). On the other hand it is clear that when *n* increases, agreement is much more difficult to obtain. An asymmetrical distribution of power of decision, amounting to institution of a mechanism of authority, would then correspond to a lower level of costs of interdependence than the democratic solution. Let us, for example, imagine that in the matter of traffic regulation, it is desirable to adopt the democratic solution which comprises the questioning of drivers, rather than the general adoption of the authoritarian solution (red lights). Such a solution would result in the interdependent costs being so high that it would soon appear to be impossible to carry it out.

Naturally, at each time that an organizational problem arises, the difficulty mentioned above is encountered, namely that the optimal type of organization cannot be determined except in extreme situations. Between these extremes, zones of uncertainty are encountered, characterized by the fact that it is impossible to determine objectively whether one solution is preferable to another. It follows that these zones of uncertainty are inevitably zones of conflict; they are associated with the formation of parties which try to improve the adoption of a particular solution. The idea of comanagement is now the password of those who believe that they could minimize the costs of interdependence by attenuating the asymmetry of the distribution of authority. The book by Buchanan and Tullock demonstrated that, although it works in certain cases, it is unreasonable to suppose that equalization of the distribution of authority always leads to an attenuation of costs of interdependence. Rousseau's idea, which stated that the costs of interdependence could be higher in a state of nature than in a state of society, may be reiterated here. Translated into modern language, this proposition states: the relationship between the distribution of authority and the level of costs of interdependence (the intensity of adverse effects) is complex and variable, subject to the cases of the figure considered. If we add that the idea of optimal distribution of authority does not generally make sense because of the existence of large zones of uncertainty, we conclude that any problem of social organization is normally a subject of conflict.

Thus, interpreted and prolonged, the analysis of Buchanan and Tullock could be approached from the viewpoint of a thesis by the German sociologist, Dahrendorf (3). According to it the conflicts of industrial societies are on the whole conflicts relative to the distribution of authority. If, as is some times done and as Dahrendorf himself indicates, this thesis be interpreted as signifying that the unequal distribution of authority is always arbitrary, unjust, intolerable, and derived from the product of violence between social groups, then the thesis is certainly false. However, Dahrendorf's thesis can also be given a generous interpretation in positive terms: except in extreme cases, it is difficult to assert that natural freedom is superior to civil freedom (and vice versa); except in extreme cases it is difficult to demonstrate that this form of organization, and particularly in the distribution of authority, is, from the viewpoint of its members, superior to any other. It follows that the problems of distribution of authority are very frequently the object of opinions and of conflicts; but this battle, with respect to the distribution of authority, does not comprise the truth of the absurd belief in the unconditional superiority of the equal distribution of authority.

The fundamental hypothesis of the study by Buchanan and Tullock is, in summary, that every organization aims to eliminate the costs which each one imposes upon another in any situation of interdependence, or in other words, the principal function of the social organization is that of the elimination of adverse effects. However, this elimination is never free; it is in turn accompanied by costs which vary, on the one hand, with the nature of the adverse effect which it seeks to eliminate, and on the other hand, with the type of organization selected. Unfortunately to the propositions of Buchanan and Tullock must be added the proposition according to which the determination of costs could not be, subject to exception, established objectively. It follows that this solution applies only on a practical basis in the opposition and confrontation of points of view. The Leibnizian ideal, _non disputemus, sed calculemus_, is inapplicable, in spite of the fallaceous title of Buchanan and Tullock's book, _The Calculus of Consent_.

I shall briefly consider the work of Schelling on segregation which furnishes another brilliant example of the relationship between adverse effects and social change (17). The fundamental result of this work is that the phenomena of social or racial segregation do not necessarily result from the segregationist attitudes. In other words, segregation could be the adverse result of behaviors which in themselves are not segregationist. To illustrate this effect of composition, Schelling has recourse to a simple model. Let us consider a checkerboard and give ourselves 20 pieces of 10 centimes and 10 pieces of one franc, symbolizing the individuals belonging, respectively, to two social or racial groups. Let us next place the 20 pieces on the checkerboard in a random manner. The resulting situation may then be interpreted as simulating the relations between the groups in a residential block. Let us then assume that the members of each of the two groups, even though no hostility or any desire for segregation with respect to the other groups exists, experience uneasiness at finding themselves in the situation of a minority. This feeling can be simulated by assuming that each of the pieces on the board, finding itself in the minority situation, will move to a position where it is surrounded by at least 50% of pieces of the same category. When this position of equilibrium is attained, it is assumed that the piece in question is satisfied and stops moving (except if, when its neighbors move likewise to attain an equilibrium, the piece finds itself again in the position of a minority). It is easy to play this game by displacing the pieces on the checkerboard. The structure of equilibrium is counterintuitive. In effect, the process produces a phenomenon of abrupt segregation. Although each of the individuals, far from experiencing hostility with respect to the other members of this group, appreciates quite well that one half of his neighbors belongs to the other group, an effect of considerable segregation is created as the pieces of one franc on the board form a sort of ghetto immersed in an environment populated by pieces of 10 centimes. Thus, the desire of each one to escape from a minority situation gives rise to the effect of a composition which caricatures the individual preferences. Although they were not in any way hostile to the one franc pieces, I have proven that as far as the 10 centimes pieces are concerned, there was a natural desire not to be in minority situation. It suffices for the welfare of the latter that half of its neighbors be of the same species as it is. This natural desire nevertheless engenders, from the moment that it is expressed by all the individuals, an undesirable effect of autoamplification: the one franc pieces will find themselves together as will the 10 centime pieces.

This fable is very important. It reveals how undesirable social states could result from an amplification, not desired by any one, of individual, natural behaviors. The elimination of these adverse effects understandably implies, aside from other possible costs, the interference of external elements with the freedom of action and of decision by individuals. An undesired and undesirable social state naturally calls for its elimination. However, the costs of its elimination could be considerable. In this case it is probable that a situation of social tension would appear, that factions will be formed, and that ideologies will run their courses. Extending Schelling's fable, a party will be formed which would plead the elimination, at all costs, of the adverse effect leading to segregation and would opt for elimination or the restriction of freedom of movement of the coins, whereas another party would proclaim the sacred character of this freedom. Briefly, we can predict that it would lead to an entrenched ideological opposition, to conflicts, and possibly to social changes.

What have we learned from this survey? I shall confine myself to several propositions which appear to me to be essential. The first is that states of social equilibrium, social tension, and, in consequence, social change could result not only in conflicts of contradictory interest but in adverse effects engendered by most of the structures of interdependence. The undesirable social states do not necessarily result from the ability of one dominant group to impose its wishes and interests on dominated groups. This game structure of _sum zero_ is not the most characteristic or the most important, although it can occasionally characterize the social life. A most important progress registered by modern sociology with respect to the explanation of social conflicts and of social change is the simultaneous discovery by a number of authors of the importance of games of _sum zero_ in the analysis of situations of interdependence. Those who continue to believe in the virture of the simplistic opposition between conflict and consensus and who conceive conflicts as necessarily having the game structure of _sum zero_ recall the physicians described by Molière.

In the second place, it has been seen that the situations of equilibrium, disequilibrium, and social tension are in some way normal. Adverse effects are omnipresent in social life. However, in a very general way, we cannot rely upon the observation of a consensus on the relative value of their uselessness in relation to the costs of their elimination. The particular case of the distribution of authority is significant from this viewpoint and the conclusions to which it leads can easily be generalized. Dahrendorf sensed this very well, but explained it poorly. A given distribution of authority cannot, under general conditions, be contested; the idea of the optimal distribution of authority is generally senseless in most undetermined cases. Since all organizational processes understandably assume that a definite distribution of authority will be adopted, this normally results in the issue of conflicts between the interested groups and will necessarily be the ground for new conflicts. As a corollary, the normal character of the ideological phenomenon will also be observed. In the absence of an objective solution (that is, considered to be optimal by all), the ideologies furnish pseudoobjective systems of criteria.

The preceding remarks show, in the third place, in a more general way, that the analysis of social change cannot resort without circumspection to analogous models borrowed from other disciplines. The analogy of the epidemiologic contagion, favored by Tarde, can be used in particular cases only. The chemical analogy, favored by Durkheim, of society-synthesis is not very useful. The biological analogy of adaptation of an alleged social organism has a changing environment; the cybernetic models, in matters of sociology, have still to prove their usefulness. The preceding analyses show that the limited interest of these analogies is due to the fact that it is difficult to analyze typical cases of social tension, social disequilibrium, and social change without considering, as a reasonable fragment, the individual with his specific capacities of motivation, occasionally of rationality, and in all cases of action. If this point is conceded it also follows that the physical analogies would not comprise more than a limited interest in sociology. Changes in social systems could not be likened to physical systems except by incongruous analogies of poor heuristic value.

As for the alleged laws of history, which were so popular among sociologists in the 19th century and still attract some sociologists of the 20th century, they could be discarded without inconvenience. No absolute spirit looks after human destiny. Of course, it is possible to observe statistical regularities and cumulative processes: the cultural phases of Sorokin are not without foundation (18); a number of phenomena assume, over time, a logistic and exponential appeal. However, these regularities do not plead in favor of the heuristic value of physical or epidemiologic analogies. For the sociologist, once he has asserted that a phenomenon presents a certain type of regularity over time, he must always explain why this is so. The explanation always results in the investigation of the individual action behind the regularities which are observed at the macrosociologic level.

## REFERENCES


1. Boudon, R.: L'Inégalité des Chances, Colin, Paris, 1973.

2. Buchanan, J.M., and Tullock, G.: The Calculus of Consent, The University of Michigan Press, Ann Arbor, 1965.

3. Dahrendorf, R.: Class and Class Conflict in Industrial Society, Stanford University Press, Stanford, 1959.

4. Downs, A.: An Economic Theory of Democracy, Harper and Brothers, New York, 1956.

5. Fararo, T., and Kosaka, L.: A mathematical analysis of Boudon's IEO model. In a symposium on l'Inégalite des Chances. Information sur les Sciences Sociales 15: 2-3, 1976.

6. Hirschman, A.O.: Exit, Voice and Loyalty, Harvard University Press, Cambridge, Mass., 1970.

7. Hotelling, H.: Stability in competition, Economic Journal 39: 41-57, 1929.

8. Lipset, S.M.: Social mobility and educational opportunity. The Public Interest, 29: 90-108, 1972.

9. Lipset, S.M., and Bendix, R.: Social Mobility in Industrial Societies, University of California Press, Berkeley, 1958.

10. Mendras, H.: La Fin des Paysans, Sedeis, Futuribles, Paris, 1967.

11. Michels, R.: Political Parties, Dover, New York, 1959.

12. OCDE: Conference on the policies of development of education, OCDE, Paris, 1972.

13. Olson, M.: The Logic of Collective Action, Harvard University Press, Cambridge, Mass., 1965.

14. Popper, K.: Misère de l'Historicisme, Plon, Paris, 1956.

15. Rousseau, J.J.: Discours sur l'Origine de l'Inégalité, In Écrits Politiques, Gallimard, Paris, pp. 166-167.

16. Schelling, T.C.: On the ecology of micromotives. The Public Interest 25: 61-98, 1971.

17. Schelling, T.C.: Dynamic models of segregation. J Math Sociol 1: 143-185, 1971.

18. Sorokin, P.A.: Social and Cultural Dynamics, Bedminster Press, New York, 1962.

19. Tarde, G.: Les Lois de l'Imitation, Alcan, Paris, 1890.

20. Thurow, L.C.: Education and economic equality. The Public Interest, 28: 66-81, 1972.

21. Touraine, A.: Production de la Société, Seuil, Paris, 1973.

22. UNESCO: Textes Choisis sur L'Économie de L'Éducation, UNESCO, Paris, 1968.

23. White, L.: Medieval Technology and Social Change, MacMillan, New York, 1930.

# *Beyond Gradient-Climbing*

Jon Elster

The aim of the present paper is largely polemical and critical. I shall argue that in spite of certain superficial analogies between the social and the biological sciences, there are fundamental differences that make it unlikely that either can have much to learn from the other. The difference, essentially, lies in the distinction between the intentional explanations used in the social sciences and the functionalist explanations that are specific to biology. Davidson (1) has argued that the attempts at psychophysical reduction are bound to hurt themselves on the stumbling-block of intentionality, and from a very different point of view a similar thesis is presented here with respect to psychobiological reduction or analogy. The argument cuts both ways: against the use of sociological methods in biology and against the transfer of biological paradigms to the social sciences. Most of the space is devoted to the first problem, because it is less extensively discussed in the literature. I argue that even if natural selection can to some extent simulate intentionality, there are important differences in the

---

This is a revised version of a paper presented at the Fourth International Congress of the International Organization for the Study of Human Development, Paris February 16-18, 1977. I am grateful to Roger D. Masters, Arthur Stinchcombe and George C. Williams for their written comments on the preliminary version.

fine grain of animal and human adaptation. Two such differences are discussed in some detail. In Parts II and III below a case is made for the capacity for global maximization being a specifically human trait not found in natural selection. In Part IV a similar argument is made concerning the specifically human capacity for strategic behavior. Part V turns to the second problem, drawing attention to some better-known difficulties of functionalist analysis in sociology. Briefly stated, my contention here is that in societies there is no general mechanism--corresponding to natural selection--that could permit us to infer that the latent functions of a structure explain the persistence of that structure.

Discussions of these problems with colleagues have shown me that my argument lends itself to two misunderstandings. First, I do not deny that interdisciplinary research, pooling the empirical and theoretical resources of biology and of sociology, can be fertile indeed. In human evolution, for example, the social structure is part of the environment that determines whether a given mutation will be beneficial or harmful. At another level, the biological components of language, intelligence, or mental illness are by now well documented, even if the relative importance of nature and nurture is a matter of controversy. My argument is directed only at the transfer of explanatory paradigms; at the appeal to intentions in biology and to functions in sociology. Second, I do not deny that cases can be found where biological adaptation leads to deviation from local maxima and to attainment of global maxima, nor that functional explanation in sociology can sometimes be useful. I am only arguing that there can be no general presumption that global maxima will be attained by natural selection; and that latent functions do not typically explain the persistence of the social structures exhibiting these functions.

We may invoke here the Kantian notion of a regulative idea, as opposed to a dogmatic one (2). A regulative idea distributes the burden of proof, in the sense of postulating which features can be assumed as a working hypothesis in the absence of specific evidence to the contrary and which features must be demonstrated in each particular case. The statement "Everything has a function" should not be taken as a dogmatic assertion about organisms (3), but rather as a statement to the effect that we may legitimately assume that a given feature has a function--i.e., could not undergo small variations without loss of reproductive capacity for the organism concerned--until the contrary has been demonstrated. The justification for this regulative idea is, of course, the general mechanism of gradient-climbing in natural selection, as explained in Part II below. On the other hand, there is nothing that could justify the regulative idea that every feature of the organism is part of a global maximum, so that _all_ variations--small or large--would imply a loss of reproductive capacity. For intentional adaptation exactly the contrary is true. Here we do have a general mechanism for attaining global maxima, so that what needs a separate explanation is rather the failure to achieve this.

As a last remark I would like to set forward what may hopefully turn out to be a self-destroying prophecy: I do not believe that my criticism, should it happen to be read, will have any great impact. The attraction of biological analogies on social scientists, in particular, seems to be so great that even the best minds are led astray. In the cabinet of horrors of scientific thought there is room, alas, not only for a Worms or a Lilienfeld, but also for a Durkheim or a Merton. At the present moment I think one can confidently predict that ethology and sociobiology between them will generate a continuous stream of pseudo-scientific papers for some years to come. I do not deny, of course, the occasional utility of biological analogies as a source for new hypotheses, no more than I would deny that some scientists may get their best ideas when reading the Bible or the _Dialectics of Nature_. I do deny, however, that biological analogies should have any privileged status. Ideas should be judged by their implications, not by their ancestors. I hope that the social scientist who proposes to spend his time on finding the social analogue of, say, dominance or flight behavior in animals (4), would first reflect upon the pseudodebates of the 19th Century concerning the correct social analogue of the cell: individual or family? There is no reason why any such analogue should exist, and that's all there is to the question.

## II. THE LOCALLY MAXIMIZING MACHINE

In this section I shall sketch a highly simplified account of the theory of natural selection. By keeping strictly to first principles I hope to avoid being too patently wrong on specific biological matters outside my competence, even if I then run the risk that the conclusions that are valid for the simplified account do not hold for the more realistic picture. I shall conceive, then, of the organisms in a population as a machine receiving inputs in the form of mutations. For simplicity we may assume that there is asexual reproduction, so that mutations are the only source of genetic novelty; alternatively we may abstract from recombination by arguing that in the long run only mutations can disturb the equilibrium (in a constant environment). The stream of inputs is random, in the sense that there is no correlation between the functional requirements of the organism and the probability of occurrence of a mutation satisfying these requirements. By mutagens such as X-Rays it is possible to increase the probability of mutations generally; by chemical mutagens it is possible to increase the probability of structurally defined subgroups of mutations, such as the probability of cytosine mutating to thymine as a result of deamination by nitrous acid; but it is never possible--and this is the central dogma of molecular biology--to modify the probability of functionally defined subgroups of mutations. (An analogy may help here. Comparing mutations to misprints, we may increase the probability of misprints by breaking the glasses of the typesetter, but there is no way of selectively increasing the probability of misprints occurring in the second edition of a book that will correct the factual errors of the first edition.) We also assume--crucially--that all mutations are small, being typically amino acid substitutions resulting from the misprint of a single letter in the genetic code.

While mutations are random, the selection process is deterministic in the sense that the machine at any given moment has well-defined criteria for accepting or rejecting any given mutation. (This means that I abstract from genetic drift and from the problem of non-Darwinian evolution.) The mutation is accepted if the first organism in which it occurs benefits in the form of higher reproductive capacity. Since the organism will leave more descendants than other organisms, the new allele will spread in the population until it is universally present. (This statement implies that I abstract from frequency-dependent selection and other sources--notably heterosis--of stable polymorphism.) One result is that the criteria for accepting or rejecting new mutations will change. The organism is now in a different state from what it was before the mutation occurred, and may be harmed or benefitted from different inputs. To put the matter briefly: the machine says Yes or No to each input according to criteria that change each time it says Yes. If the machine ever arrives at a state in which it says No to each of the (finitely many) possible inputs, we say that it has reached a local maximum. The population climbs along a fitness-gradient until it reaches a point from which all further movement is downward only; and there it comes to a halt. For a given initial state, several local maxima may be accessible, the choice between which depends upon the random order in which the mutations happen to occur.

A further exploration of the logic of this <u>locally maximizing</u> <u>machine</u> must take account of the possibility of environmental changes. If the environment changes, the criteria for saying Yes or No to mutations will also change. A mutation is not beneficial or harmful in itself, only with respect to a given genetic background (itself the result of previously accepted mutations) and a given environment. With a changing environment it may very well be the case that even instantaneous local maxima are never attained, if the organism cannot keep pace with its surroundings. The notion of an "environment," however, is ambiguous. In the first place, environmental changes may be changes in the geological or climatic conditions to the extent that they cause evolutionary change without being themselves affected by it. In the second place, some parts of the environment are themselves evolving organisms or the effects of such evolution. If a population is constantly subject to exogenous environmental change, a steady state can never be attained, but to the extent that the environment is itself made up of (or the effect of) evolving organisms, it makes sense to ask whether a *general equilibrium*--where all organisms have attained local maxima relatively to each other--can be realized. The conditions for the existence and attainability of a general equilibrium would presumably be very complex. Here I only stress the general idea that evolution may be seen under the increasingly complex assumptions of a parametric environment, a strategic environment, and an exogenously changing environment. In my abstract model I deal with the first case only, although I believe that the conclusions are easily extended to the second case. With the third case, however, entirely different considerations become relevant. In particular the phenomenon of preadaptation--stemming from the fact that "any structure has properties beyond those for which it was constructed" (5)--permits the attainment of global optima that in a constant environment would have been inaccessible. This, however, is a chance phenomenon only; not a general mechanism comparable to natural selection.

Enough should have been said by now to explain why the realization of local maxima in the organic world is not a matter of course (as it is in the model). There is no logical objection to the idea of a world where the rate of change of the environment relatively to the mutation rate was so high that most organisms most of the time were badly adapted to each other and to the inorganic environment. In the world we know, however, the infinitely subtle adaptations found in the structure and behavior of organisms are a fact that for millenia have evoked the wonder and (with less justification) the admiration of naturalists. These adaptations do indeed seem to simulate intentional design. In many well-documented cases the natural solution to structural and functional problems is strikingly close to the solution that would have been chosen by the engineer or the economist working on the same problem. In some cases animals and men _are_ facing the same problems, such as building, communicating, or flying, so that the actual solutions can be compared. As shown by d'Arcy Wentworth Thompson in his classical work _On Growth and Form_ as well as by several recent authors (6), these solutions are often strongly convergent. In recent ecological work (7) "Nature" is seen as an economist rather than as an engineer; optimal budgeting, linear programming, profit maximization, cost minimization, and efficient allocation are now as much a part of evolutionary theory as of economics.

It is sometimes said that the theory of natural selection is a tautology; that the survival of the fittest means the survival of the survivors. If by fitness we mean genetic fitness, this is true enough. If, however, we understand fitness as ecological adaptation, as measured for example by the life span of the organism, the survival of the fittest is turned into an empirical proposition to which there are many important counter-examples. To take but one example, natural selection works so as to produce the optimal sacrifice of parents for offspring: too much sacrifice for a given offspring may reduce the chance of having more offspring later on, whereas too little sacrifice would damage the "investment" already made in the offspring. If, on the other hand, the aim were to maximize ecological adaptation, zero sacrifice would clearly be the optimal strategy; indeed the best strategy would be to have no reproduction at all. The confusion between ecological and genetic fitness occurs in various forms in the literature. A rather crass example is found in a recent work on the philosophy of biology, where the author states that "the process of child-bearing in humans" shows that "what may be useful to the group (e.g., a species) may have no utility" (8). A more subtle error (or potential for error) underlies the frequent use of the term "parental investment" in recent evolutionary theory (9). Whereas economic investment means the sacrifice of something (i.e., consumption) now in order to get more _of the same thing_ later on, parental investment means the sacrifice of something (i.e., ecological fitness) in order to get _more of another thing_ (i.e., genetic fitness).

## III. THE GLOBALLY MAXIMIZING MACHINE

For the purposes of the present argument the crucial features of the locally maximizing machine refer to what it _cannot_ do. In particular the machine is incapable of _waiting_ and of using _indirect strategies_. These notions are defined as follows. The machine is capable of waiting if it can say No to a favorable mutation in order to be able to say Yes to an even more favorable one later on. Suppose that of a given protein form B is superior to A and C is superior to B, and that because of the structure of the genetic code one-step mutations are possible from A to B and from A to C, but not from B to C. If the mutation to B should occur first, the organism is incapable of saying No. The machine is capable of indirect strategies if it can say Yes to an unfavorable mutation in order to be able later on to say Yes to an extremely favorable one. Suppose that A is superior to B and C to A, and that one-step mutations can occur from A to B and from B to C, but not from A to C. If B comes about, the organism is incapable of saying Yes. As a gradient-climber the machine has its myopic eye fixed to the ground, and is incapable of taking account of what happens beyond the next hill. As one author puts it, "future events or possibilities simply cannot have any effect on the gene pools of organisms" (10). It is no doubt true that "even mutations to codons that are completely synonymous with each other may in many instances alter the future mutational possibilities of the genotype" (11), but these possibilities cannot set up a selectional pressure now that could give an edge to an apparently neutral mutation over the status quo.

On the other hand waiting and indirect strategies are crucial features of human choice, for which reason man may be seen as a globally maximizing machine. This characterization goes back to Leibniz, who saw man as being uniquely capable of "reculer pour mieux sauter" (12). Some examples of global maximization shall be given from economic and political life:

| | Waiting | Indirect strategies |
|---|---|---|
| Politics | Anti-activism | Anti-reformism |
| Economics | Patent system | Investment |

A few comments may be in order. Investment is perhaps the simplest example of global maximization that requires bypassing a local maximum; one step backwards in order to take two steps forward. As observed above, parental investment in the animal reign is not really an example of investment in this sense. If, for example, one generation accepted a less-than-maximal number of offspring for the sake of a larger number of grandchildren, this would indeed be investment in the strict sense, but no such idea has ever been entertained in the literature (as far as I know). The patent system has the paradoxical feature that "by slowing down the diffusion of technical progress it ensures that there will be more progress to diffuse" (13). As Joseph Schumpeter repeatedly stressed (14), the maximal exploitation of present possibilities often is an obstacle to the maximal creation of new possibilities. As suggested by Maurice Meisner (15), this proposition may also be applied activism as an obstacle to long-term action. A revolutionary class must be capable of waiting or marking time, of saying No to favorable opportunities. Conversely political reformism may be seen as the refusal of indirect strategies. With Karl Popper's "piecemeal social engineering" (16) only local maxima will ever be reached, whereas a revolutionary class is capable of saying Yes to unfavorable possibilities. [It is interesting to observe that the neglect of the distinction between local and global maxima is also found in Popper's work on biology, where he wrongly asserts that "What Darwin showed us was that the mechanism of natural selection. . .can simulate rational human action directed towards a purpose or aim" (17)]. These observations are purely formal and do not imply any particular political view. I am only saying that the human <u>capacity</u> for revolutionary behavior--for marking time and using indirect strategies--is a feature that distinguishes man from the activitist-cum-reformist (i.e., opportunist) behavior of natural selection.

A further difference, of course, is that man can increase the probability of functionally defined subgroups of "mutations" (i.e., new ideas, inventions or institutions). As Michael Ruse put it, "one can talk of many, if not most, new cultural elements appearing by virtue of the fact that they are needed" (18). We are not constrained to sit back and wait for the right mutation to occur; we can actively seek out the new kind of invention or institution that is needed. The relative importance of randomness and directedness in technological development is a matter of some controversy, and at least one recent author

(19) has argued that random inventions can generate a process that simulates consciously directed search. His argument, however, is intended to explain technological change in the 19th Century, and does not pretend at universal validity. It seems obvious, indeed, that the modern large corporation very actively searches out new techniques with highly specific properties, in contradistinction to the more myopic learning-by-doing that may have characterized the small firm of the last century. At this point we may also observe that even if random search is observed to take place in human choices, this does not mean that the search is generated by a random process. On the contrary, a prior calculation may have found a random search to be the most rational procedure, perhaps because a consciously directed search would involve unacceptably high costs of acquiring and processing information (20).

The last remark brings us to a new level of problems, and requires some distinctions that will turn out to be crucial for the analysis. The reader, indeed, may justifiably feel somewhat confused at this point, for exactly which are the phenomena under comparison. If we are contrasting animal and human adaptation, the appropriate comparison would presumably be between human and animal behavior, whereas we have been staging a confrontation between human choice and natural selection. Actually three different phenomena are involved: animal adaptation, human adaptation, and the natural selection that has fashioned them both. An instructive example for bringing out the relation between the three is the theory of pursuit games. In the theory of differential games there is a standard example called "the homicidal chauffeur game," where an automobile on an infinite unobstructed parking lot attempts to run down a pedestrian (21). The pursuer has a high speed, but poorer mobility because of the larger radius of curvature of the car. It is intuitively clear, and can be proved rigorously, that if initially the pursuer is very close to the evader, the optimal strategy for the pursuer is to drive away from the evader, and then, when far enough, turn around for a direct pursuit. One step backward is required for two steps forward; a typical example of the indirect strategies that man is capable of using and actually does use in many real-life pursuits.

In the animal reign such pursuits are found at two levels. At the individual level we can see the fox pursuing the hare across the fields any autumn day; at the species level the fox is hunting the hare across the generations, continually adapting to the counter-adaptations of the latter. I shall have a few words to say about the species pursuit, and then dwell somewhat more at length on the individual pursuit. The upshot of Part II above was that in the species pursuit, indirect strategies and waiting are impossible (in the simple model), for an organism that took one step backwards would not leave any descendants to take the two steps forward later on, whereas an organism that refused to take one step forward is inconceivable. (In reality, of course, genetic drift (22) and preadaptation invalidate this proposition, but both of these are essentially chance phenomena). The crucial question then becomes whether the local maximization of natural selection is capable of producing a maximizer; whether the gradient-climbing process at the species level is capable of evolving individuals that can raise their eyes from the ground and look beyond the nearest hill. In an abstract sense we know that the answer must be positive, since man himself is both a product of natural selection and a deployer of indirect pursuit strategies. It may be useful, however, to take a closer look at the problem as it arises in the predator-prey context.

Edmunds (23) and Curio (24) have studied in some detail the behavior of prey and predators respectively. Prey, it seems, do not use indirect strategies, but are quite capable of waiting; predators are capable of both. For example, even if flight may reduce the present danger, it may increase the danger at later times by signalling the existence of the prey to the predator, for which reason the prey has "learned" (through selection) not to panic (25).

A symmetrical example of waiting in predators is the "speculative and unobtrusive trot which brings the lion closer to a prey than any speedy, and hence warning run" (26). Also "a variety of animals are able to make detours while stalking" (27) and to intercept the flight path of the prey (28). The prey, in turn, have evolved a counter-strategy to interception. This is not the direct counterstrategy of a continuous change in direction so as to always move away from the pursurer; rather the prey uses the so-called "protean display," a randomly changing direction that confuses the pursurer without depending upon the path of the latter (29). According to Curio "relatively few predators have overcome the protean defense" (30), but one of these few must be the cheetah cited by Curio himself, which uses a less-than-maximal speed in pursuing its victim because the "deceleration allows the cat to follow each twist in the unpredictable route of its quarry" (31). This looks like the following scenario. In a first stage, predator and prey both follow the simple strategy of direct pursuit and evasion. In a second stage, the predator evolves the indirect strategy of path interception. In a third stage, the prey evolves the protean counter-strategy; and in a last stage, the predator responds with the waiting behavior of suboptimal pursuit speed. I do not know whether the actual evolutionary process was anything like this, but at any rate the component stages are all well-documented. The main point of the preceding paragraph is brought home by the observation that in this sequence each step is indeed advantageous to the species that takes it.

This digression into prey-predation behavior shows, first, that animals are capable of globally maximizing behavior and, second, that such behavior is found in highly specific and sterotyped situations. For the present purpose the second idea is the crucial one. The characteristic feature of man is not a programmed ability to use indirect strategies or adopt waiting behavior in specific situations, but rather a generalized capacity for global maximization that applies even to qualitatively new situations. Given the facts of pleiotropy, genetic drift, and preadaptation it is not very surprising, perhaps, that one-step mutations can produce two-step strategies. It is much harder at present to see how the step from the situation-specific to the generalized capacity for global maximization can have occurred. The breakthrough may, perhaps, be observed in "Imo the monkey genius" who, when given a mixture of sand and wheat, invented the indirect strategy of first throwing the mixture on the water (one step backwards) and then retrieving the lighter wheat grains (two steps forward). Edward Wilson described this behavior as "a qualitatively new element: throwing the food temporarily away and waiting a short period before retrieving it" (32). I suggest that this qualitative leap is at least the use of an indirect strategy that is atypical (even if not unknown) in the animal realm, and perhaps even a deployment of a generalized capacity for indirect strategies. To say that man is a "globally maximizing machine" is to stress the generalized capacity; to him global maximizing comes as easily as does local maximizing to natural selection. By contrast the situation-specific cases of global maximization evolved through natural selection are accidental only.

To sum up we may say that in creating man, natural selection has transcended itself. The leap implies precisely a leap from nonintentional to intentional adaptation. In cases of situation-specific indirect strategies or waiting there is no need to appeal to intentionality. Someone might want to contest this for the predator-prey examples given above, but at any rate there can be no element of intentionality in the behavior of "a cell which can, as it were 'see' several moves ahead, and make an energetically unfavorable move if by so doing it arrives at a position from which it can make an exceedingly favorable one in a subsequent transition" (33). By contrast the use of globally maximizing strategies in novel situations must imply an analysis of the situation, a scanning of several possible moves and the deliberate choice among them. "La sélection s'opère, non parmi les possibles, mais parmi les existants" wrote Francois Jacob (34) in a phrase that was the point of departure for the reflections offered in the present paper. Man, by contrast, can choose among unactualized possibles.

## IV. STRATEGIC BEHAVIOR IN ANIMAL AND HUMAN SOCIETIES.

In Part II above a brief reference was made to a distinction between a parametric and a strategic environment. In this Part it will be argued that natural selection always acts as if the environment is parametric, even when it really is a strategic one, whereas man is capable of taking the strategic nature of his environment into account. (In the first reference I dealt with the relation between a population and the other populations that make up its environments, whereas here I deal with the relation between an individual and the other individuals in the population that makes up his environment. The logical point, however, is the same.) My point can also be stated by saying that man--in contrast to natural selection--is capable of realizing the solution to games without a dominant strategy for any of the actors.

The parametrically rational actor treats his environment as a constant, whereas the strategically rational actor takes account of the fact that the environment is made up of other actors and that he is part of their environment, and that they know this, etc. In a community of parametrically rational actors each will believe himself to be the only one whose behavior is variable; acting upon these inconsistent beliefs the actors will generate unintended and perverse consequences of the kind explored by social scientists from Marx to Keynes or from Mandeville to Boudon (35). Such behavior is irrational from a collective point of view, even if individually each actor satisfies the criteria of rationality. A necessary (but not sufficient) condition for collective rationality is the transition to strategic thinking. In the strategic or game-theoretic mode of interaction each actor has to take account of the intention of all other actors, including the fact that their intentions are based upon their expectations concerning his own. This looks like an infinite regress of the "I think that he thinks that I think. . . ." variety, but actually this is not the case. Rather than giving the formal definition of a solution in the game-theoretic sense, I show below by some examples how the actors may bypass the infinite regress and converge upon a predictable course of action. From the present point of view the important thing to note is that human actors not only make their choices on the basis of expectations about the future, as pointed out at the end of Part III; they also sometimes _decide on the basis of their expectations about the expectations of others_. This is the true Leibnizian monad, each individual reflecting the totality from his point of view. The transparency and symmetry of this interaction ensures that the fate of the actors is in their own hand, whereas a community of parametrically rational actors will be in the grip of causal forces that elude them and that perpetually make their plans come to nought. At any rate strategic actors will escape what we may call supra-intentional causal forces; the problem of subintentional (largely biological) causality is a darker one (36). I shall now argue that there is no such strategic interaction in the animal reign, and that the neglect of this difference between the social and the biological fields has led a number of recent authors down a blind alley. To set out the distinction in its simplest form, we shall look at two games that will be solved differently in a human society, but identically in animal societies. The first is the Prisoner's Dilemma, that has been extensively studied by psychologists, economists and political scientists (37). I shall specify the structure of this game in a somewhat unusual form, where it appears as a game between "me" and "all others." We let variables x and y range over the set of strategies (C,D), where "C" stands for Cooperation and "D" for Defection. The expression "/x,y/" then means that "I choose x, while all the others choose y." The Prisoner's Dilemma is then defined by the fact that for each action (i.e., "me") the relevant alternatives are ranked in the following order of preference: 1) /D,C/; 2) /C,C/; 3) /D,D/; and 4) /C,D/. It is immediately seen that whatever the others do, my best strategy is always D. As the others are in the same position, they will also choose D, the ensuing and fully predictable outcome being /D,D/, which is worse for everyone than /C,C/. This shows why strategic interaction is not a sufficient condition for collective rationality.

Among the innumerable examples of this structure from the social sciences, we may single out the problem of work motivation. In an egalitarian society without solidarity between the workers no one will put any effort into the work even if it is better for everyone if everyone does so. For a biological example, we may turn to Williams' explanation of the evolution of schooling in fishes (38). Letting "D" stand for the tendency to seek towards the middle of the school and "C" for the absence of any such tendency, Williams argued that we have in fact a Prisoner's Dilemma where schooling evolves because it is individually favorable even if collectively it may lead to the extinction of the population. *Ceteris paribus* it is always better to be in the middle of the school than at the outskirts, because the fish in the middle is less exposed to predators. If, however, everyone seeks towards the middle, the school as a whole becomes more compact and more exposed to predators. A simpler example is provided by aggression: in a population of peaceful organisms an aggressive mutant will get the upper hand, but its descendants may peck each other to death.

The second game, to be compared with the Prisoner's Dilemma, is less well known. It has been extensively studied by Amartya K. Sen under the heading of the "Assurance Game" (39), and differs from the Prisoner's Dilemma in the ranking of the first two alternatives: 1) /C,C/; 2) /D,C/; 3) /D,D/; and 4) /C,D/. This game differs from the Prisoner's Dilemma in the crucial feature that there is no dominant strategy, i.e., no strategy that is the best whatever the others do. How should a rational actor behave in this game? With less than perfect information, he should use the maximum-criterion and choose the strategy that guarantees him the highest minimum payoff. In the Assurance Game this is clearly the strategy D, for the worst that can happen to him with this choice is alternative 3, whereas the worst that can happen should he choose C is alternative 4. If on the other hand we assume perfect information for all players, then they will all choose C in the confident expectation that everyone else will do the same, as no one has anything to gain from doing differently. In the Prisoner's Dilemma the requirement of full information is not necessary for the solution to emerge, but in the Assurance Game the slightest uncertainty or suspicion will make an actor choose D rather than C. Only in a fully transparent situation will the actors converge upon the collectively rational behavior. The difference between the two games may also be expressed by saying that in the Prisoner's Dilemma the optimum /C,C/ is both individually inaccessible (no one will take the first step towards it) and individually unstable (everyone will take the first step away from it), whereas in the Assurance Game it is individually inaccessible and individually stable. Because human actors are able to act upon expectations about expectations, the stability of the optimum will make them converge towards it in spite of the individual inaccessibility.

For a sociological example of the Assurance Game, we may again take the problem of work motivation, where Sen has argued that in a socialist society preferences should reflect the Assurance Game rather than the Prisoner's Dilemma. Hard work effort would not (or so I should hope) be a dominant strategy under socialism; no one would prefer hard (and in most cases monotonous) work for its own sake, regardless of what the others are doing. Still solidarity with the others could bring about the required effort: if I am certain that the others take their share of the work, I prefer taking mine rather than being a free rider.

It is an empirical problem to determine whether the Assurance Game is often solved, in the sense of /C,C/ emerging as a result of individual calculations. Presumably the very stringent information requirements can only be fulfilled if the information is supplied from the outside (e.g., by the government) or if the community is sufficiently small so that everyone can really come to know everyone else. In the present context, however, we are less concerned with the _probability_ of the cooperative behavior than with the very _possibility_ of mutually dependent cooperative actions. In human societies this possibility can be realized; in natural selection it cannot. I shall argue that a not uncommon fallacy with biologists is to think that _stability explains evolution_, so that the emergence of the solution in a game without dominant strategies can be explained by its immunity to free riders (40). Individual accessibility is required for the solution to be realized in biological games, and individual stability can only assure its maintenance once evolved (41). I now proceed to some examples of this fallacy.

In the otherwise admirable book by Edward Wilson on _Sociobiology_ we may take our first example of the fallacious reasoning from stability to emergence. In the chapter on "The prime movers of social evolution" (an expression that definitely suggests mechanisms for the _emergence_ rather than just the maintenance of behavior) we find the following explanation of schooling in fishes:

> One potential variation on the selfish herd strategy is the utilization of a 'protector' that consumes part of the population but

> more than compensates by ecluding other predators. The widespread coral fish *Pempheris oualensis* forms schools of a few hundred or thousand individuals that find shelter during the day in well-shaded holes, coral passages and caves facing the open sea. They share these hiding places with one or a few kinds of predatory fishes, mostly the serranid *Cephalopholis argus*, which feed on them in limited amounts. Since the predators are territorial, the *Pempheris* gain to some extent by schooling and thus restricting their exposure during the daytime to only one or a few of their enemies. By jointly saturating the favored predators with more than they can consume, the individual members of the school are favored with an increased probability of survival (42).

I read this text as *stating* that the sharing of hiding places with a predator is an individually stable adaptation, and as *implying* that no further explanation is required for the emergence of this behavior. The adaptation is stable because any single fish that tried itself in the open sea would be more exposed to predation than it is in the caves. On the other hand this does not prove that the adaptation is individually accessible (e.g., a dominant strategy) and indeed the "safety in numbers" through joint saturation of the predators cannot have favored the fish that first ventured into the caves. There must be some threshold beyond which there is safety in numbers; and the evolution up to this threshold cannot be explained by invoking retroactive causality or anticipations (43). A similar ambiguity is found in Curio's discussion of "swamping the appetite of predators" as a general counterstrategy of the prey(44).

A second example concerns the use of the notion of *mixed strategies* in animal and human games. A mixed strategy is a probability distribution, fixing for each pure strategy the probability that it will be chosen. The revolution of game theory stems from von Neumann's 1928 proof that certain apparently insoluble games admit of a solution in mixed strategies. Now a genuine mixed strategy (involving nonzero probabilities of at least two pure strategies) can never be a dominant strategy. For social games this has the implication that realization of the solution requires perfect information; for animal games that the solution can only be realized by chance. Let us explore these two propositions by some examples.

Among the very few attempts to explain real-life problems through the notion of mixed strategies the recent work of Boudon on education (45) and on relative deprivation (46) may be cited. In his work on the latter problem, he tries to explain the classical and paradoxical finding that the level of frustration in the army was highest where the objective possibilities of promotion were largest. Imagine, for simplicity, a cohort of 20 MP and another cohort of 20 Air Force members. The first belongs to a group characterized by small possibilities for promotion and low level of frustration, whereas in the second both promotion potential and frustration are high. Let us further assume that in order to participate in the "promotion game" you must stake 1 franc's equivalent of effort. The promoted receive the equivalent of 5 francs in renumeration. Five of 20 AF and 2 of 20 MP will achieve this, all participants having the same chance of success. It is then easy to see that all the 20 AF will stake 1 franc, with 15 out of 20 being disappointed, whereas only 10 MP will enter the game, of which 8 are necessarily disappointed. The reason why 10 MP only enter the game is, of course, that each of the 20 finds it individually rational to use the mixed strategy of entering the game with probability 1/2. This ingenious and elegant demonstration suffers from the defect that the use of the mixed strategy is rational only if each individual is certain that the others will do the same, which requires a quite implausible amount of information about the other players.

Still this defect is an empirical one only, and there is nothing intrinsically impossible in the model. If human beings always were as rational as they can sometimes be, this is how they would behave. No such general justification can be given for the realization of the solution in animal games requiring the use of mixed strategies. Actually mixed strategies are even worse off in this respect than nondominant pure strategies. The latter may emerge as solutions to a game where they _are_ dominant strategies, and then may be maintained even when the game is transformed (through environmental change) into a game where they are individually stable without being individually accessible. This cannot happen with mixed strategies, because they cannot be dominant in any game whatsoever. For an example where mixed strategies are used in evolutionary theory, we may take "the logic of animal conflict" as analyzed by John Maynard-Smith (47) [or rather the simplified version of his analysis presented by Richard Dawkins (48)]. We here assume that the population consists of two genotypes, hawks and doves, these being not the biological species, but two behavioral variants of the same species. When a dove meets a dove, each of them has a 50% chance of winning 50 "evolutionary units" and 50% of losing 10 units. When a hawk meets a hawk, each of them has a 50% chance of winning 50 and 50% of losing 100. When a hawk meets a dove, the first wins 50 and the second loses 10 with certainty. It is then easy to show that the proportion of doves to hawks in the population will converge towards 5:7, i.e., a stable polymorphism. So far, so good. The crucial and illegitimate step comes when Maynard-Smith and Dawkins both affirm that a similar situation can evolve without polymorphism, namely if each individual uses the dove strategy with probability 5/12 and the hawk strategy with probability 7/12. This would be true in a population of human actors with perfect information, but not for animal actors incapable of evolving a strategy that is optimal only against the same strategy chosen by everyone else.

For a third example of the fallacy we may turn to a recent article by Gary Becker (49), who argues that the notions of egoism and altruism are more subtle and intertwined than has usually been thought. Roughly the argument is that an egoist has a clear incentive to act as an altruist, because an increase in the income of the altruist will rebound upon the egoist. A crucial assumption is, of course, that the egoist is capable of correctly anticipating the altruist's action. For the egoist, renouncing of an increase in income that is brought about at the expense of an even larger reduction in income for the altruist, could not possibly be a dominant strategy; it is effective only against altruists and not against other egoists. Among the many interesting conclusions of Becker's analysis is that an altruist may actually be better off than an egoist, because the "beneficial indirect effects on the behavior of others may dominate the direct 'disadvantages' of being altruistic" (50). Becker then applies the same scheme to the problem of genetic fitness, where he claims that the appeal to group selection or kin selection in order to explain altruistic behavior is less indispensable than has been thought by most biologists. If altruist genes confer individual benefits on their bearers because others take account of them in order to maximize *their* benefits, then ordinary natural selection at the individual level is sufficient to explain the emergence of altruism.

Now this is without any doubt a truly brilliant argument, but I would like to question the biological half of it. Let us look at a population of egoists where a mutant altruist first appears. The egoists, of course, will treat him as if he were just another egoist, because they have no way of "knowing" or "anticipating" that the altruist will confer upon them some of the benefits they confer upon him. For this reason the altruist gene will quickly disappear from the population. In other words, Becker seems to have made the mistake of imputing to lower organisms the faculty of acting upon expectations that is specific to man. And *if* the organisms had this faculty, I submit (tentatively) that they would rather use it in the following manner: "Why should we reciprocate towards the altruist, since this would make us worse off relatively to him even if in absolute terms we should all improve our lot?" What is maximized by natural selection is differential fitness, not absolute fitness.

As a last remark I would like to observe that even in the animal reign game-theoretic solutions <u>may</u> emerge in the absence of dominant strategies, viz., in the case where successive adapations to the adaptations of others converge towards the solution (as in a damped cobweb cycle). A fascinating example of this possibility is provided by W.D. Hamilton in his game-theoretic analysis of "Extraordinary Sex Ratios." He studies the case of double parasitism (two parasitoids on one host); more precisely the situation where there are two types of females each of which is trying to maximize its genetic fitness through selection of a sex-ratio (ratio of males to total progeny) that is optimal relatively to the ratio chosen by the other. Arguing first that fitness is proportional to the number of inseminations by sons + the number of daughters, he then shows that for a female using sex-ratio x on a host with another female using ration $x_o$, the fitness will be proportional to

$$\frac{x}{x + x_o} [(1-x) + (1-x_o)] + (1-x).$$

For the female using $x_o$ the fitness will be proportional to the same expression, with $x_o$ and x changing place everywhere. If we now restrict the set of possible ratios to (0,1/4, 1/2, 3/4, 1)--a simplification that makes no difference to the analysis--the following payoff-matrix emerges:

| | | $x_o$ | | | | |
|---|---|---|---|---|---|---|
| | | 0 | 1/4 | 1/2 | 3/4 | 1 |
| | 0 | ? | 1,2.5 | 1,2 | 1,1.15 | 1,1 |
| | 1/4 | 2.5,1 | 1.5,1.5 | 1.2,1.3 | 1,1 | 0.9,0.6 |
| x | 1/2 | 2,1 | 1.3,1.2 | 1,1 | 0.8,0.7 | 0.7,0.3 |
| | 3/4 | 1.5,1 | 1,1 | 0.7,0.8 | 0.5,0.5 | 0.4,0.1 |
| | 1 | 1,1 | 0.6,0.9 | 0.3,0.7 | 0.1,0.4 | 0,0 |

Disregarding the undetermined (and biologically uninteresting) case of both females choosing to produce no males, we immediately see that the only equilibrium point (and thus the solution) of the matrix is when both players use 1/4. We also see that the ration 1/4 is not a dominant strategy, for if x chooses 1, $x_o$ will choose 0 rather than 1/4 and conversely. Nevertheless Hamilton is justified in saying that "through trial and error, two naive players would quickly learn that constant playing of 1/4 was the optimum-yielding strategy " (51). If, namely $x_o$ starts out with one of the set (/,1/4,1/2,3/4), then 1/4 is x's best answer, and in the next round 1/4 will also be the best counter from $x_o$. If $x_o$ begins with 1, x will counter with 0, to which $x_o$ will reply 1/4, which in turn will induce x to choose 1/4. Thus whichever initial strategy is chosen by $x_o$, the sequence of replies and counterreplies will make the players converge upon the solution in a maximum of four steps. This example shows that the presence of a dominant strategy is not a necessary condition for the game-theoretic solution to emerge, but nevertheless the point remains valid that in the absence of specific arguments to the contrary, there is no presumption that the solution will be realized in a game without dominant strategies. Hamilton offers such an argument; Wilson, Maynard-Smith and Becker do not. Their procedure seems to rest upon unjustified assumptions concerning the burden of proof; they take for granted what must be demonstrated.

## V. FUNCTIONALIST EXPLANATION IN SOCIOLOGY

In this part I shall turn to more familiar matters that, nevertheless, seem worthwhile reconsidering. I shall argue that functionalist explanation in sociology rests upon an ill-conceived analogy from biology, and that a closer analysis of purported functionalist explanations shows that in virtually all cases one or more of the defining features are lacking. Now there is no general agreement as to which features should enter into the definition of a functional explanation, and to the reader whose definition differs from mine the following may seem beside the point. Still I think that my use of the term is in fairly good agreement with the standard expositions by Merton (52) and Stinchcombe (53), so that I am not setting up a straw man to knock down. On my definition, then, _an institution or a behavioral pattern X is explained by its function Y for group Z if and only if_:

1. Y is an _effect_ of X
2. Y is _beneficial_ for Z
3. Y is _unintended_ by the actors producing X
4. Y (or at least the X-Y causal relation) is _unrecognized_ in Z
5. Y maintains X by a causal _feedback_ loop passing through Z.

In this paradigm the actors involved in X may be identical with the group Z, or there may be a smaller or greater degree of overlap. Now for some comments upon the criteria. That functions are a subspecies of effects is, I think, relatively uncontroversial. To say that an institution "has a function" must imply that it _does_ something, i.e., generates some effect. The notion of a beneficial effect is rather more obscure. I shall take it to mean that Y is a local maximum of some state variable of which the actors in Z always want more rather than less. It would be unduly restrictive to require that Y be a global maximum, for then the biological paradigm would break down at the outset. Needless to say, the actors producing X may be engaged in an activity of global maximization (of some other variable). That Y be unintended and unrecognized is part and parcel of the Mertonian paradigm of latent functions, to which I am here restricting myself. By contrast it is rather hard, but not impossible, to find the last criterion in Merton's account of functionalism (54). On the other hand a causal feedback loop is made _the_ defining feature of Stinchcombe's account.

I shall now argue for a series of propositions. First, it is (nearly) impossible to find any cases of functional analysis where the presence of all the features 1-5 is demonstrated. Second, there exists a naive brand of functional analysis that from the presence of features 1, 3, and 4 concludes to the presence of feature 2 and perhaps also to feature 5. Third, there exists a more sophisticated, but equally fallacious, brand of functionalism that from the presence of features 1-4 concludes to the presence of feature 5. Fourth, there is a brand of (especially Marxist) functionalism that argues--fallaciously--that an institution can be maintained by its _long-term_ effects if these satisfy criteria 2-4, even if the short-term effects are negative. Fifth--turning now to more positive statements--analyses satisfying criteria 1-3 and _either_ criterion 4 _or_ criterion 5 are of fundamental importance indeed in the social sciences.

Starting with the last proposition, accounts satisfying requirements 1, 2, 3, and 5 may be called _filter-explanations_ whereas analyses satisfying criteria 1-4 may be dubbed _invisible-hand explanations_ (55). [Following Boudon (56), theories satisfying criteria 1, 3, and 4, but not criterion 2, may be called _malign-fist explanations_]. Militarily financed research may be analyzed by a filter-explanation: if academic personnel apply for military funds in order to be able to conduct the research that they would have done in any case (i.e., with money from any other source), the Department of Defense may serve as a filter that selects some applications and rejects others (57). The resulting composition of research will be beneficial to the military interests, while wholly unintended by the individual scientist, who can argue truthfully that no one has told him what to do. This might also be called a case of _artificial selection_, where the feedback loop _operates through the recognized effects of the structure that is to be explained_. Merton's account of the "Matthew Effect" (58) in science may serve as an example of invisible-hand explanations: here X = a disproportionate amount of recognition to senior scientists in case of co-authorship and simultaneous discoveries, Y = more rapid dissemination of important results, and Z = the scientific community. Another example of invisible-hand explanations is the case where X = labor-saving inventions, Y = reduction of the wage level, and Z = the community of capitalist entrepreneurs (59).

I know of only one example where all features 1-5 are actually present, viz., in the attempt of the Chicago school of economists to explain profit-maximizing as a result of the "natural selection" of firms by the market (60). The anomaly that led to this attempt was the following. On the one hand the observed external behavior (factor combinations, output level, etc.) seemed to indicate that firms in general adopt a profit-maximizing behavior. On the other hand the internal decision-making process of the firm did not seem to be guided by this objective; rather some rough-and-ready rules of thumb were found to be typical. To bridge this gap between the output of the black box and its internal workings, the economists in question postulated that some firms just happen to use profit-maximizing rules of thumb and others not; that the former survive whereas the latter go extinct; that the profit-maximizing routines tend to spread in the population of firms, either by imitation or (which is assumed here) by takeovers. If, therefore, we set X = a certain rule of thumb, Y = profit-maximizing, and Z = the set of firms, we have a paradigm of functional analysis. This argument clearly is modeled on a biological analogy, and it works [to the extent that it does work (61)] only because the notions of fitness, survival, reproduction and inheritance can be transferred without too great modifications.

In this case the feedback loop is explicitly demonstrated. In most cases, however, it is postulated rather than demonstrated. Everything seems to point to an implicit regulative idea used by sociologists: if you can demonstrate that a given structure has some unintended, unrecognized and beneficial effects, then you have also explained why it exists and persists. That is, the existence of the feedback loop can be taken for granted in the absence of direct evidence to the contrary. Merton, for example, when referring to the "function" of multiple discoveries for science (62), must presumably be interpreted as stating that these are maintained by their unintended, unrecognized and beneficial effects. The benefits are discussed in some detail; that they are unintended and unrecognized seems plausible, but the feedback loop is not even sketched. Perhaps Merton is just offering an invisible-hand explanation, but to this reader at any rate his whole approach seems to tend at an <u>explanation</u> of the structure in question. I believe that a large and indeed vast number of similar cases could be cited, where the analysis is acceptable and interesting if read as an invisible-hand explanation, while the terminology and conceptualization point to something substantially stronger.

Stinchcombe is among the relatively few writers who have recognized that "the nature of reverse causal links from consequences back to structure [is] a sore point in functional theory" (63). He also sketches six distinct mechanisms that could fill this gap. The first is the natural selection in the literal (biological sense). The second and (I think) the sixth invoke "natural selection" of firms by markets. The third relies on conscious planning, and is really out of place in this context.

The fifth seems to be quite similar to what was called above filter-mechanisms. The fourth is explained in the following terms: "Without planning, people may find consequences of behavior satisfying. Thus church services might be maintained without much planning to achieve theological ends, because people find the social interaction, or the respectability, satisfying" (64). This Durkheimian example has been discussed in some detail by I.C. Jarvie, who concludes that the functionalist analysis has no explanatory power, because people's "reasons for going to church may have very little to do with what they are doing in going to church" (65). I tend to agree with this conclusion, which means that the only successful example is the one I have singled out above, viz., selection of firms by the market. In biology both of the following may be taken as regulative ideas:

- Every persisting structure has a function (i.e., has beneficial effects through which it is maintained)
- Every persisting structure with beneficial effects has a function (more specifically, is maintained through these effects).

Were these propositions to be read as general laws, the second would be redundant, but the logic of regulative ideas is different. A regulative idea admits of counterexamples and is stronger, the fewer the counterexamples. As the class of counterexamples to the second proposition is smaller than the class of counterexamples to the first (66), the second is worth stating in its own right. The above discussion of Merton and Stinchcombe concluded that in the social sciences there can be no regulative idea analogous to the second of these propositions; no general presumption (with the burden of proof on him who holds the opposite view) that the relevant subclass of beneficial effects maintain their causes. <u>A fortiori</u> this must hold for the sociological analogy to the first regulative idea, which is justly criticized by Merton (67). We may point out at this point that in contemporary sociology vulgar Marxism is the best example of this fallacy; there does indeed seem to be an implicit regulative idea of this school that "Every institution or behavioral pattern in capitalist society serves the interest of capitalism and is maintained because it serves this interest." To this fallacy vulgar Marxism adds another, which has to do with the dazzling manipulation of the time perspective. If, say, the actions of the State go counter to short-term capitalist interests, this has the function of safeguarding long-term capitalist interests. (Heads I win, Tails you lose.) Now this is not only an arbitrary procedure (68), but a theoretically inconsistent one, because functional analysis cannot invoke indirect strategies, as explained above. To the extent that the State is maintained through the effects of its actions for the capitalist class, the negative short-term effects should make it disappear (or change) before the long-term positive effects come to be felt; unless, of course, some conspiracy theory is substituted for the functionalist explanation.

## NOTES

1. Cp. especially "The Material Mind" in P. Suppes *et al*. (Eds.), *Logic, Methodology and Philosophy of Science IV*, Amsterdam: North Holland 1973. The characteristic feature of Davidson's view is that he is simultaneously materialist, determinist, *and* antireductionist. The following passage brings out the essence of his view: "If a certain psychological concept applies to one event and not to another, there must be a difference describable in physical terms. But it does not follow that there is a single physically describable difference that distinguishes any two events that differ in a given psychological respect." (*Op.cit*, p. 717). That is, to conclude from determinism to reductionism is to commit the same fallacy as when from the fact that everything has a cause it is concluded that there is something which is the cause of everything.

2. *Kritik der reinen Vernunft*, B 670 ff. (A 642 ff.); *Kritik der Urteilskraft* para. 70 ff. The simplest example is perhaps Kant's reinterpretation of the "principle of continuity" from a dogmatic assertion to the effect that the set of organic forms can be ordered in a continuous (or dense?) series, to a regulative principle that exhorts us to seek, between any two forms, an intermediate one. The statement "Everything has a cause" is in Kantian philosophy a constitutive principle to which there can be no exceptions, whereas the idea that "Everything has a function" is a regulative one that does admit exceptions.

3. For an extreme statement of this (non-non-Darwinian) view, we may take the following: "A 'functionless site' is simply one the function of which has not yet been determined." (E. Mayr, *Populations Species and Evolution*, Cambridge, Mass.: Harvard University Press 1970, p. 127). A statement along the lines sketched in the present approach is the following: "The experimental

study of adaptation has unravelled adaptive values in such unobstrusive and inconspicuous details of organismic organization that one should think of a character as having survival value unti the contrary has been demonstrated." (E. Curio, "Towards a Methodology of Teleonomy," _Experientia_ 1973, p. 1049). Cp. also note 66 below.

4. The work of Roger D. Masters is an example of this approach; illustrated in his "Politics as a Biological Phenomenon," _Social Science Information_ 1975 and his attempt to find biological analogies to Albert Hirschman's categories of "exit," "voice," and "loyalty," in _Social Science Information_, forthcoming.

5. S.N. Salthe, _Evolutionary Biology_, New York: Holt, Rinehart and Winston, 1972, p. 9.

6. E.G. Leigh, _Adaptation and Diversity_, San Francisco: Freeman & Cooper 1971, part I; cp. also S.J. Gould, "D'Arcy Thompson and the Science of Form," in M. Greene and E. Mendelsohn (Eds.), _Topics in the Philosophy of Biology_, Dordrecht: Reidel, 1976.

7. For surveys, see M. Cody, "Optimization in Ecology," _Science 183 (1974)_ and D.J. Rapport and J.E. Turner, "Economic Models in Ecology," _Science 195 (1977)_.

8. M. Simon, _The Matter of Life_, New Haven: Yale University Press 1971, p. 82.

9. Cp. especially R.L. Trivers, "Parental Investment and Sexual Selection," in B. Campbell (Ed.), _Sexual Selection and the Descent of Man_, London: Heinemann, 1972. He defines "parental investment as _any investment by the parent in an individual offspring that increases the offspring's chance of surviving (and hence reproductive success) at the cost of the parent's ability to invest in other offspring (p. 139). Investment in the economic_ sense would rather be "any investment by

the parent in an individual offspring that increases the offspring's chance of having surviving offspring (and hence reproductive success) at the cost of the parent's chance of having surviving offspring."

10. Salthe, _op.cit_., p. 133.

11. G.L. Stebbins and R.C. Lewontin, "Comparative Evolution at the Level of Molecules, Organisms and Population," in _Proceedings of the Sixth Berkeley Symposium on Mathematical Statistics and Probability_, vol. V, Berkeley and Los Angeles: University of California Press 1972, p. 24.

12. Leibniz, Die Philosophische Schriften, Ed. Gerhardt, vol. III, p. 346, p. 578, vol. VII, p. 568. These references actually invoke indirect strategies as a part of the divine rationality when constructing the best of the possible worlds. Similar propositions about man are the following: "Les appétitions sont comme la tendance de la pierre qui va le pluis droit mais non past toujours le meilleur chemin vers le centre de la terre, ne _pouvant pas prévoir_ qu'elle rencontrera des rochers où elle se briscera, au lieu qu'elle se serait approchée davantage de son but, si elle avait eu _l'esprit et le moyen de se détourner_.... Le bonheur est donc pour ainsi dire un chemin par des plaisirs; et le plaisir n'est qu'un pas et un advancement vers le bonheur, le plus court qui se peut faire _selon les présentes impressions_, mais non pas toujours le meilleur. On peut manquer le vrai chemin, en voulant suivre le plus court, comme la pierre allant droit, peut recontrer trop tôt des obstacles, qui l'empêchent d'advancer assez vers le centre de la terre. Ce qui fait connaître, que c'est la raison et la volonté qui nous mènent vers le bonheur, mais que le sentiment et l'appétit ne nous portent que vers le plaisir." (_Ibid_., vol. V, p. 175, p. 182, italics added). See also my _Leibniz et la Formation de l'Esprit Capitaliste_, Paris: Aubier-Montaigne 1975, Ch. VI, for a further discussion.

13. Joan Robinson, _The Accumulation of Capital_, London: Macmillan 1956, p. 87.

14. _Captialism, Socialism and Democracy_, London: Allen and Unwin 1954, p. 83.

15. _Li Ta-Chao and the Origins of Chinese Communism_, Cambridge, Mass. Harvard University Press 1967, p. 166 ff.

16. _The Poverty of Historicism_, London: Routledge and Kegan Paul 1957 p. 64 ff. Popper contrasts the piecemeal approach with the Utopian approach which insists that the society must be remodelled as a whole. He never explicitly states that only the steps that lead to short-term improvements are acceptable on the piecemeal approach, but several passages (cp. especially p. 58) strongly suggest that this is indeed his opinion. It should perhaps be added that there _are_ good arguments for being rather cautious in the use of indirect strategies in politics. In the first place our factual knowledge about what will happen two steps ahead is considerably more shaky than our short-term predictive power, and in the second place our preferences may change (endogenously) during the traverse. For the second point, see my "Note on Hysteresis in the Social Sciences," _Synthese_ 1976.

17. _Objective Knowledge_, Oxford: Oxford University Press 1972, p. 267.

18. "Cultural Evolution," _Theory and Decision_ 1974, p. 432.

19. P. David, _Technical Choice, Innovation and Economic Growth_, Cambridge: Cambridge University Press 1975, p. 61 ff.

20. Economists will have recognized the theory of "satisfying as maximizing" in this phrase; for some problems created by this theory see note 36 below.

21. Cp. R. Isaacs, _Differential Games_, New York: Wiley 1965 and O. Hajek _Pursuit Games_, New York: Academic Press 1975.

22. Sewall Wright ("Random Drift and the Shifting Balance Theory of Evolution," in K. Kojima (Ed.), _Mathematical Topics in Population Genetics_, Berlin: Springer 1970, p. 12 ff.) has explained how random drift may carry a population over the "saddle" of local maxima. Conversely Edward C. Wilson (_Sociobiology_, Cambridge, Mass.: Harvard University Press 1975, p. 24) mentions that genetic drift may prevent the species from ever reaching a _local_ maximum "if the species perches on a knife ridge leading up the adaptive peak."

23. _Defence in Animals_, London: Longman, 1974.

24. _Ethology of Predation_, Berlin: Springe, 1976.

25. Edmunds, _op.cit._, p. 251.

26. Curio, _op. cit._, p. 133.

27. _Ibid._, p. 137.

28. _Ibid._, p. 148 ff.

29. Edmunds, _op.cit._, p. 145-146, Curio, _op.cit._, p. 153-154.

30. Curio, _op.cit._, p. 153.

31. _Ibid._, p. 147, quote slightly modified.

32. _Op.cit._, p. 171.

33. R. Rosen, "Morphogenesis," in R. Rosen (Ed.), _Foundations of Mathematical Biology_, vol. II, New York: Academic Press 1972, p. 55-56.

34. _La Logique du Vivant_, Paris: Gallimard 1970, p. 313.

35. I discuss this more extensively in my *Logic and Society*, forthcoming.

36. By subintentional causality I refer to the processes that lead to preference formation and to the endogenous change of preferences. To be rational is to choose according to consistent preferences, but do we choose our preferences? It is certainly possible to do so but then the problem presumably arises at one further remove. A fully transparent rationality would have to be able to *choose itself* a problem involving self-reference in much the same way as "the optimization whose scope covers all considerations, including its own costs" (S.G. Winter, "Optimization and Evolution," in R.H. Day and T. Groves (Eds.), *Adaptive Economic Models*, New York: Academic Press 1975, p. 56.

37. Cp. especially A. Rapoport and A. Chammah, *Prisoner's Dilemma*, Ann Arbor: University of Michigan Press 1965; M. Olson, Jr., *The Logic of Collective Action*, Cambridge, Mass.: Harvard University Press 1963; M. Taylor, *Anarchy and Cooperation*, London: Wiley 1976.

38. G.C. Williams, *Adapatation and Natural Selection*, Princeton: Princeton University Press 1965, p. 212 ff.

39. Cp. especially his "Isolation, Assurance and the Social Rate of Discount," *Quarterly Journal of Economics* 1967 and *On Economic Inequality*, Oxford: Oxford University Press 1973.

40. This conclusion was reached independently by Michael Taylor, in his "Altruism and Cooperation: Their Evolution by Natural Selection and their Role in Primitive Societies," Mimeo: University of Essex, 1976, especially p. 40: "The problem. . . is that, although it is reasonable to suppose that an equilibrium, once reached, will be maintained, there are not the reasons for supposing that it will be attained in the first place

which there are in the case of a game played by rational human agents."

41. In social games an important category are the games (such as "Chicken") where the cooperative behavior is individually accessible (i.e, everyone finds it in his interest to adopt the cooperative behavior if the others do not) and individually unstable (everyone prefers to use the noncooperative strategy if the others cooperate). Such games often do not have a (noncooperative) solution, which means that the behavior of the players cannot be predicted on the sole assumption of their rationality; additional psychological assumptions are required. In the corresponding biological games such preference structures will lead to a stable polymorphism, through frequency-dependent selection.

42. _Op. cit._, p. 38.

43. Elsewhere Wilson recognizes this. In this discussion of Robert Trivers' theory of reciprocal altruism he remarks that "The machines for bringing the gene frequency up to the critical value must lie outside the game itself." (_Op.cit._, p. (121).

44. Curio, _op.cit._, p. 33-34.

45. Boudon, Cibois et Lagneau, "Enseignement Superieur Court et les Pieges de l'Action Collective," _Revue Francaise de Sociologie 1975_, especially note 41. For a discussion of the need for mixed strategies in the case discussed in note 41, see my "Boudon, Education and the Theory of Games," _Social Science Information_ 1976.

46. R. Boudon, "La Logique de la Frustration Relative," Paper presented to the ECPR Workshop on Political Theory, Louvain 1976.

47. "The Logic of Animal Conflict," _Nature (246)_, 1973: "The Theory of Games and the Evolution

of Animal Conflict," *Journal of Theoretical Biology 1974*.

48. *The Selfish Gene*, Oxford: Oxford University Press 1976, Ch. 5.

49. "Altruism, Egoism and Genetic Fitness," *Journal of Economic Literature 1976*.

50. *Ibid*., p. 820.

51. *Science (156), 1967*, p. 486.

52. *Social Theory and Social Structure*, Glencoe, Ill.: The Free Press 1957, Ch. I.

53. *Constructing Social Theories*, New York: Harcourt, Brace and World 1968, Ch. 3.

54. In the explicit paradigm Merton (*op.cit*., p. 50 ff.) lists eleven dimensions of the analysis; the feedback loop is not among them. But in his more informal discussion Merton several times implies that the ascription of (latent) functions to some pattern explains the existence and persistence of the pattern. This, indeed, is the central idea of the section (*op.cit*., p. 64 ff.) where Merton explains that the distinction between manifest and latent functions "clarifies the analysis of seemingly irrational social patterns." The most explicit statement (with a rare occurrence of the word "explain") is found in a reinterpretation of Veblen's *Theory of the Leisure Class*: "among these latent functions, which help explain the persistence and the social location of the pattern of conspicuous consumption, is its symbolization of 'pecuniary strength and so of gaining or retaining a good name'." (*Op.cit*., p. 69).

55. I borrow this terminology from R. Nozick, *Anarchy, State and Utopia*, Oxford: Blackwell 1974, p. 118 ff., though I use it to express a

distinction that does not quite coincide with his.

56. "La Main et le Poing," Contrepoint 1975.

57. Here I draw upon S.A. Glantz and N.V. Silbers, "Department of Defence R & D in the University," Science 186, 1974.

58. R. Merton, The Sociology of Science, Chicago: University of Chicago 1973, Ch. 20.

59. Cp. especially W. Fellner, "Two Propositions in the Theory of Induced Innovations," Economic Journal 1961 and P. David, op. cit., p. 33 ff. (In my Logic and Society this example is discussed at length. In this invisible-hand explanation no one has made the fallacy of concluding to the existenceof some feedback loop, but rather the fallacy of attributing to individual entrepreneurs the intention to mechanize production in order to reduce wages, which is, of course, unthinkable in a competitive economy where the actions of individual entrepreneurs have an infinitesimal impact on the wage level. That is, from the fact that everyone is observed to behave in a certain way and that it is to the benefit of everyone that everyone behaves in this way, one has concluded (fallaciously) that everyone behaves in this way in order to obtain this benefit.

60. For brillant expositions of this theory, see S.G. Winter, "Economic 'natural selection' and the Theory of the Firm," Yale Economic Essays 1964 and "Optimization and Evolution," op.cit.

61. In addition to the objections cited by Winter in the articles referred to in the preceeding note, we may add the local-global difficulty; unless we assume that all possible rules of thumb are represented in the population of firms, there is no presumption that a global maximum shall be attained by this mechanism.

62. Merton, *The Sociology of Science*, Ch. 17.

63. *Op.cit.*, p. 85.

64. *Ibid.*, p. 86.

65. "Limits to Functionalism and Alternatives to it in Anthroplogy" in R.A. Manners and D. Kaplan (Eds.), *Theory in Anthropology*, London: Routledge and Kegan Paul 1968, p. 198-199.

66. Counterexamples to the first propositions can be found at four distinct levels. At the most fundamental level, one should not think of mutations as "having the function" of generating the evolutionary process. Natural selection is not a result of natural selection; (G.C. Williams, *op.cit.*, p. 128) mutations require causal, not functional explanations. At an intermediate level the theory of non-Darwinian evolution argues that the number of functionless sites (or rather site-holders) in proteins is very large indeed; at the more molar level pleiotropy is a general mechanism for generating and maintaining neutral or even harmful features. The counterexamples of the fourth level are also counterexamples to the second proposition: these refer to beneficial effects that arise as the statistical summation of many individual adaptations. G.C. Williams refers to the misconception "that when one demonstrates that a certain biological process produces a benefit, one has demonstrated *the* function, or at least *a* function of the process" (*op.cit.*, p. 209). This fallacy is formally identical (with teleonomy replacing teleology) to the fallacy discussed in note 59 above.

67. *Social Theory and Social Structure*, p. 30. ff.

68. "Notoriously any argument can be turned to any effect by juggling with the time scale." (M. Brooke, "Problems in the Decision-Making Process,"

in M. Brooke and H. Remmers (Eds.), *The Multinational Company in Europe*, London: Longman, 1972, p.93.

# *Sociological Theories of Social Change*

Rolf Ziegler

From its beginnings sociology has been preoccupied with the study of social change. Yet this area remains one of the most ambiguous and loosely defined in the discipline. Many of the nineteenth century predecessors of modern sociology were heavily influenced by the biological sciences.

Classical evolutionary theory described the history of human societies as a unilinear process of development from a "backward" to an "advanced" state usually through a series of stages of ever increasing complexity. Compte (3) postulated that all societies passed through three stages from the first Theological epoch, when speculative thinking dominated, through a transitional Metaphysical period, to the final Scientific and Industrial stage in which positivism displaces religious and metaphysical imagination. Other conceptualizations of the evolutionary process are of the "ideal-type" form: According to Maine (12) social order changed from patriarchal kinship bonds based on ascribed status to relationships based on contract--two polar types of society foreshadowing and influencing the famous dichotomizaion of Tönnies (24) between Gemeinschaft (community based on common moral sentiments) and Gesellschaft (association based on rational self-interest and contractual relationships).

Society was seen by these scholars as a social organism with different structures specializing in different aspects of survival. As environmental pressure increases and survival becomes precarious either because resources are inadequate to increasing population or because of conflict with other societies, the less differentiated and, therefore, less efficient structures divide into functionally more specialized ones. The increase of structural differentiation makes integration extremely problematic and calls for special integrative mechanisms. The change from what Spencer (22) called "incoherent homogeneity to coherent heterogeneity" necessitates a governing agency which developes more and more into a hierarchically organized administrative authority.

Evolutionary thinking has been criticized for its vague concept of "prime movers" of history, its lack of empirical testing, and its emphasis on an unidirectional, inevitable, and unalterable course of development; nevertheless, this kind of analysis has undergone a real renaissance in recent years. The neo-evolutionists point to the long-run trend of continuing growth in population; continuing advances in technology, production, and capital accumulation; and to the increasing complexity and differentiation of social systems. They, of course, acknowledge exceptions to this general trend but insist that these deviations do not disprove a trend with respect to central tendencies of the frequency distributions. Regarding the explanation of this basic pattern the newer evolutionism is increasingly materialistic and considers the following to be the main determinants (10): 1) man's genetic heritage, 2) the technologies he slowly fashioned to enhance this heritage, 3) the environmental barriers to human activity and technological advance, specifically environments that impede the flow of information or preclude the adoption of more subsistence technologies, and 4) the competition and war between societies striving to maintain their territorial and resources bases.

At first glance it may seem strange to look for parallels between evolutionism and conflict-theoretical approaches like Marxism; actually, however, there are at least two strong similarities. Both describe human history as progressing toward enhanced dominance over the natural environment and consider technology, or in Marx's terminology, the "productive forces," the driving force behind this process. But it is well-known, Marx thought of this development not as a smooth, continuously changing process but as a succession of different stages brought about as the final revolutionary result of an ongoing struggle between different social classes. Each stage is associated with a dominant class of owners of capital goods [or according to Dahenrendorf's (5) generalization with a class of people holding authority positions] who claim a disproportionate share of society's production for themselves. The ruling class's struggle for preserving and enhancing their privileged position not only sharpens the conflict with the dominated class but progressively becomes an impediment to the evolving technological base. Finally, there ensues a revolution, installing a different ruling class and transforming the social structure into a technologically suitable production system.

Functionalism is in many respects the twentieth century legacy of early evolutionary theory, but its emphasis shifted from the analysis of the developmental process to the study of the conditions for internal stability of societies interacting with an uncertain environment. As indicated above, structural differentiation does endanger social cohesion. Mere functional interdependence, as Durkheim (6) argued, does not insure *solidarité* which requires the institutionalization and internalization of an integrated system of norms, which in turn is legitimated by a set of shared values. Functionalists impute to social systems homeostatic capabilities enabling them to restore equilibrium once the latter is upset. Sources of structural change are either endogeneous strains, i.e., unbalances between different subsystems, or exogeneous challenges leading to "adaptive upgrading" (17) through differentiation of the previous more diffuse structure.

"Rise and fall" theories constitute a fourth approach in the study of social change. These theories emphasize cyclical patterns rather than cumulative ones. They view development as the unfolding of inherent potentialities analogous to organismic patterns of birth, growth, maturity, decline, and death (21).

In this presentation (15,18) we are neither able nor do we intend to present these different approaches in full detail; rather we wish to make clear the strategy involved in constructing such theories. All of them try to describe and explain global patterns of development by what might be called a holistic approach that either neglects the microlevel of individual or small decision units altogether or at least analyzes the relation between the micro- and macrolevels insufficiently. If the main concern of theories of social change is how men react to conditions of their own making and in doing so change these conditions we have to deal with this relation between the two levels. As Hernes (8) has aptly stated in a recent article: "Social change is mediated through individual actors. Hence theories of structural change must show how macrovariables affect individual motives and choices and how these choices in turn change the macrovariables." He added two other *desiderata* of theories of social change which also point to shortcomings of the traditional approaches: "First, we should be able to use the same basic approach in studying both constancy and change, implying that a stable structure should be seen as a process in (temporary) equilibrium. It also means that the theory must include an explanation of why the parameters governing the process are themselves regenerated or not changed." It is not sufficient to postulate a tendency toward equilibrium as done by functionalists or to deny it as their critics do. The existence of equilibria and their stability which show up in the time-trajectories of the variables describing the system, are not properties to be postulated in addition to the mechanisms linking the rates of change of these variables to their state at time *t* or any earlier period but are logical consequences of these causal mechanisms. Finally, Hernes added, "We should be able to incorporate intrinsic sources of change into our model of a system. Changes in social systems may be induced from the outside, but not all changes can be thus induced."

Hernes pointed out that it is useful to distinguish between three levels of structural change. There is the output structure, e.g., the composition of a population by age and sex, which is generated by a birth and death process, whose structure can be specified in a set of equations. This generating process is governed by a set of parameters, the age-specific birth and death rates, which may be called its parameter-structure. Now it is well known that if, in the example given, the process- and parameter-structures remain constant the output-structure, i.e., the population pyramid, will move toward an equilibrium and will return to it after being disturbed; but of course, the time-trajectories of other output-structures need not move toward equilibrium. They may oscillate as in the cobweb example given below. While a change of output-structure will probably correspond to what Parsons called "processes within the system," changes of either parameter- or process-structure could be considered as "processes of change <u>of</u> the system" (16). Demographic transition from high to low fertility and mortality rates going through intermediate stages of first a reduction of death rates followed after a lag by a fall of the birth rate may serve as a familiar example of a change of parameters. Finally, the whole structure of the generating process may change if, e.g., emigration starts after the discovery of a new continent.

Boudon has emphasized the need for the kind of sociological models just described: the observed regularities on the macrolevel are reduced to individual choices in a context where macrophenomena act as constraints and incentives and are themselves influenced by aggregates of individual actions or of collective decisions. He showed that it is neither necessary to subscribe to a theory of conspiracy and trace back all undesirable results to a sum zero conflict of interests nor that common interests automatically lead to collective action.

We want to demonstrate with a few more examples the usefulness of such kinds of models. The first example describes the dynamics in the college job market using the classic cobweb feedback system well-known to economists. American studies show a pattern of recurrent shortages and surpluses of certain types of college-trained specialists (e.g., in engineering, accounting, physics, and business administration) in years past. These cyclical movements can effectively be explained by the following underlying lagged feedback structure (7): Young people are highly sensitive in their educational and career decisions to salaries and job opportunities but because of intrinsic lag in "educational production" the supply of graduates is determined by market conditions several periods earlier. For example, if salaries in a field are high many students are attracted to it, but they are graduated four or five years later. Unless demand has increased commensurately they will "oversupply" the market, depressing salaries and employment opportunities and cause, thereby, a decline of enrollment leading again to a shortage another four or five years later. This cyclical pattern is especially pronounced in specialties in which there is a close one-to-one tie between education and occupation. This example has demonstrated the possibility of fluctuations of output-structure, given constant parameter- and process-structures.

Next we will illustrate a process which transforms random fluctuations of exogeneous disturbances into a dampened and wave-like pattern of behavior.

The cohort, defined as "the aggregate of individuals (within some population defined) who experienced the same event within the same time interval," (17) has repeatedly been proposed as a useful concept in the study of social change. The differential impact of historical events upon birth cohorts during their formative periods of life allows us to speak, for instance, of a war-or a post-war generation; but the differential change--proneness--with increasing age people become on the average less likely to change (13)--has important consequences for the behavior of different age-cohorts and of the total population. It can be shown that the differential change-proneness makes the response to a series of events smoother, dampens it, and introduces a wave-like pattern, even though the underlying stimuli change erratically.

> The smoothing effect will be stronger, and the oscillatory pattern, the tendency to form 'swings,' more pronounced for advanced ages, since the behavioral series is then based on a larger number of terms in the stimulus series. The behavior of older people will lie closer to a long-term average and be less liable to show violent fluctuations. This makes adjacent cohorts behave more alike in later years. . . . Smooth oscillations will also appear in a time series in a time series of observations, each based on the behavior of the entire active population, that is, across cohorts . . . the response series, visually presented in a graph, will show a succession of gradually rising and falling 'tides,' giving a much stronger impression of order than the stimulus conditions would if similarly presented (2).

Our final example deals with phenomena involving sudden large variations caused by an underlying, continuous, and smoothly changing process. We consider the influence that changes in the opinion of the supporters of a government have upon the policy adopted by the administration (19). The relevant segment of the population may be the whole constituency under democratic rule or that part of the ruling elite upon whom the leaders depend for continuous support under any other form of government. The leaders adopt that policy for which they get maximal support and are supported by those whose own opinion is close to the official policy. To be more specific let us assume the case of a nation deciding upon its level of military action, ranging from surrender or withdrawal to all-out war. Let us further assume that the public opinion is in favor of moderate action which then will be the policy adopted.

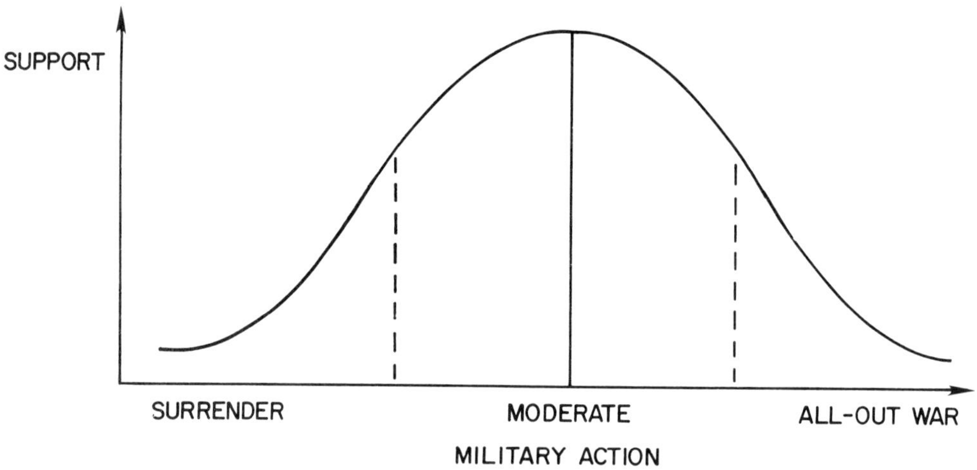


When public opinion changes we suppose that the leaders try to maximize locally the support for their policy, i.e., they will move in that direction which increases the support but will stop moving if support is decreasing. The adaptation will occur instantaneously or at least much faster than the public opinion is changing.

If the distribution now gets skewed toward less military action, administrative policy will change toward the doves (from $x_3$ to $x_1$). It may happen that for reasons we will analyze in a moment a process of polarization starts and that the group of hawks becomes stronger and stronger, but the official policy doesn't change very much because following this trend would first reduce public support. Finally, there is a sudden jump (from $x_8$ to $x_9$) when, technically speaking, the local maximum of the doves disappears. If the process reverses the leaders will first stick to the hawks, delaying the sudden jump toward the doves (from $x_{13}$ to $x_{14}$) until the local maximum of the hawks disappears. In both directions the "catastrophic change" is delayed and will occur at different distributions of public support. What are the causes of the polari-

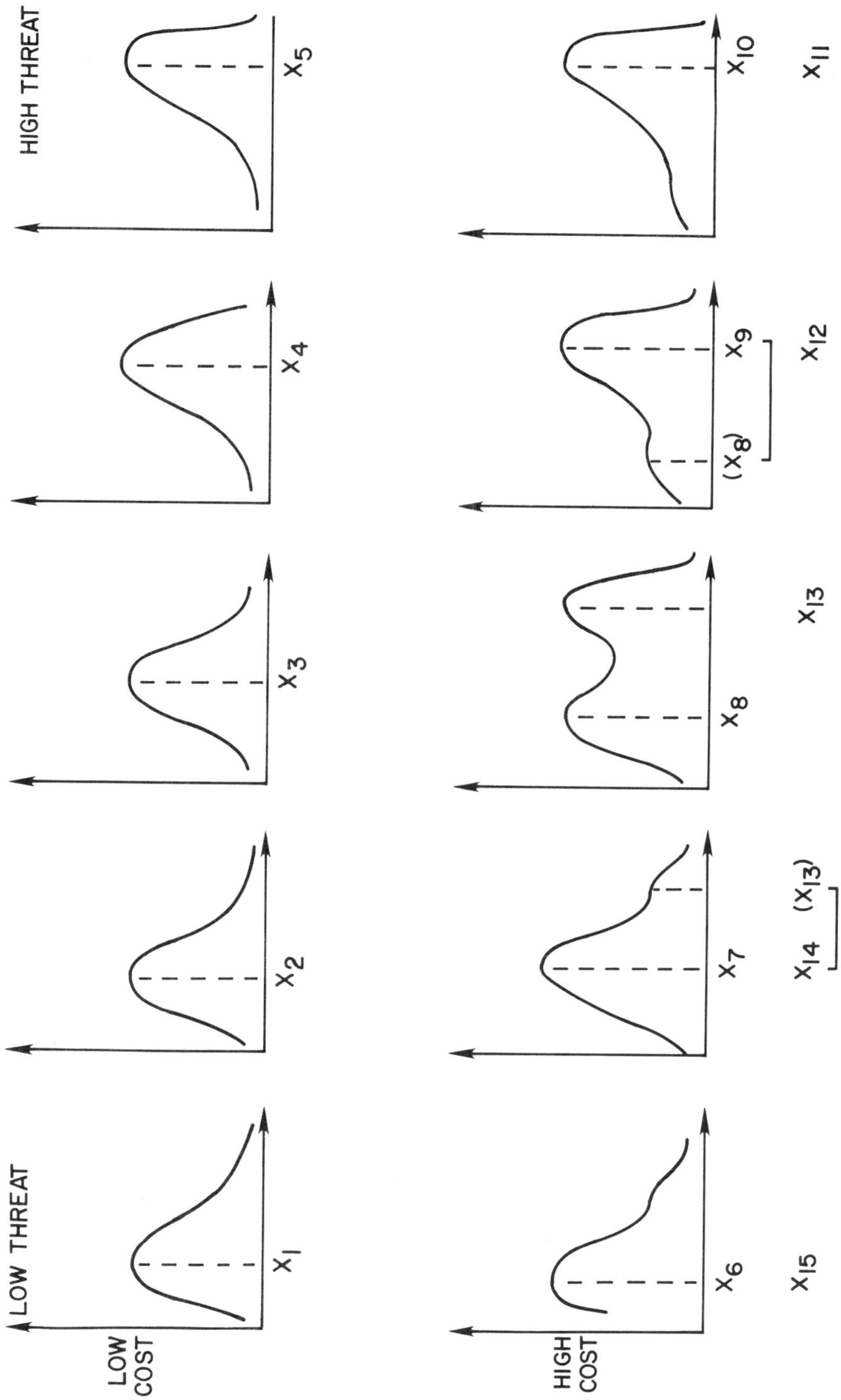
HIGH THREAT
LOW THREAT
LOW COST
HIGH COST
X5
X4
X3
X2
X1
X10
X9
(X8)
X8
X7
X6
X11
X12
X13
X14
(X13)
X15

zation of opinion? They are the combined result of two factors: threat, which pushes the public opinion in favor of strong military action, and cost, which splits the population into doves and hawks.

Summing up, we may describe the relationship between threat and cost as the independent factors, and the strength of military action as the dependent variable as follows: when the cost is low, military action will be strengthened continuously with increasing threat and vice versa. When the cost is high and the initial threat low, increasing it will for a long time apparently have little impact on the policy but suddenly will lead to the catastrophic change to a hawk policy. If the threat now lowers at further high cost the leaders will remain entrenched in their war-like policy until they suddenly abandon it.

We cannot go further into the model and the underlying mathematical theory, called "catastrophe-theory," which has been developed by the French mathematician, Rene Thom, and has recently been applied to phenomena involving sudden large variations in biology, physics, and the social sciences [here, e.g., the behavior of stock exchanges or prison disturbances (25)].

We want once more to stress the feature of this multi-level model which in our discussion is the most important. On the level of the population we have a smoothly changing opinion distribution which, by the dynamics described, produces delayed and sudden changes in policy at the administration level.

In conclusion, I shall remark that in the social sciences we should not be content to describe the global regularities and trends of societal development but should always search for the underlying individual actions in order to explain--and here I am repeating myself--how men react to conditions of their own making and in doing so change these conditions.

## REFERENCES


1. Applebaum, R.P.: Theories of Social Change, Chicago, 1970.

2. Carlsson, G., and Karlsson K.: Age, cohorts and the generation of generations. Am Soc Rev 35: 715, 1970.

3. Comte, A.: Plan des Travaux Nécessaires pour Organizer la Société, Paris, 1822.

4. Comte, A.: Cours de Philosophie Positive, Paris, 1830, p. 42.

5. Dahrendorf, R.: Class and Class Conflict in Industrial Society, Stanford, 1958.

6. Durkheim, E.: De la Division du Travail Social, Paris, 1893.

7. Freeman, R.: The Overeducated American, New York, 1976.

8. Hernes, G.: Structural change in social processes. Am J Soc 82: 513-547, 1976.

9. Isnard, C.A., and Zeeman, E.C.: Some models from catastrophe theory in the social sciences. In Collins, L. (Ed.), The Use of Models in the Social Sciences, London, 1976, pp. 44-100.

10. Lenski, G.E.: Social structure in evolutionary perspective. In Blau, P.M., (Ed.), Approaches to The Study of Social Structure, New York and London, 1975, pp. 135-153.

11. Lenski, G.E.: History and social change. Am J Soc 82: 548-564, 1976.

12. Maine, H.S.: Ancient Law, London, 1885, 5th edition.

13. Mannheim, K.: The problem of generations. In _Essays on the Sociology of Knowledge_, New York, 1952, pp. 276-322.

14. Moller, H.: Youth as a force in the modern world. _Comparative Studies in Society and History 10: 237-260, 1967-68_.

15. Moore, W.E.: _Social Change_, Englewood Cliffs, N.J., 1963.

16. Parsons, T.: _The Social System_, Glencoe, Ill., 1951, p. 481.

17. Parsons, T.: _Societies: Evolutionary and Comparative Perspectives_, Englewood Cliffs, N.J., 1976, p. 22.

18. Ponsionen, J.A.: _The Analysis of Social Change Reconsidered_, The Hague, 1965.

19. Ryder, N.B.: The cohort as a concept in the study of social change. _Am Soc Rev 30: 843-861, 1965_.

20. Smith, A.D.: _The Concept of Social Change_, London, 1973.

21. Sorokin, P.: _Social and Cultural Dynamics_, New York, 1937, p. 41.

22. Spencer, H.: _Principles of Sociology_, London, 1876, p. 90.

23. Swanson, G.E.: _Social Change_, Glenview, Ill., 1971.

24. Tönnies, F.: _Gemeinschaft und Gesellschaft_, Leipzig, 1887.

25. Zeeman, E.C.: On unstable behavior of stock exchanges. _J Math Econ 1: 39-49, 1974_.

26. Zeeman, E.C.: Prison disturbances. (Mimeo).

# *The Organization of the Eukaryotic Genome*

Giorgio Bernardi

Our understanding of the organization of the prokaryotic genome and of the regulation of its expression can be considered as fairly satisfactory. In contrast, the organization of genes in the eukaryotic genome is one of the major open problems in today's molecular biology. This issue can hardly be overestimated, since it is of capital importance for understanding evolution, differentiation, and oncogenesis.

We will attempt here, first, to introduce the major questions concerning the organization of eukaryotic genomes; second, to summarize the picture emerging from the experimental approach which has been most widely used in recent years, namely kinetic studies of DNA renaturation; and, third, to present a new approach developed in our laboratory along with the results obtained so far.

## INTRODUCTION

A comparison of prokaryotic and eukaryotic cells reveals that the major differences concern the size, the structure, and the organization of the genome.

### Genome Size

The amount of (nuclear) DNA per eukaryotic (haploid) cell (genome size), is constant in each species and covers an extremely wide range of values. Some fungi have genome sizes practically equal to those of bacteria, whereas some animals and plants have genome sizes 10,000 times as large. A closer look at available data indicates that wide variations of genome sizes are found within single phyla, within single genera, and even within interbreeding species. It is unlikely that these differences imply corresponding differences in the amount of genetic information. If these differences are, therefore, neglected, and only the minimum genome sizes found in each phylum or order are considered, a ratio of about 1,000 is found between the smallest genome size of prokaryotes and the largest genome size found in eukaryotes, namely the genome size of mammals; furthermore, a well-defined trend for the minimum genome size to increase with evolution is evident.

## Genome Structure

The genetic material of eukaryotes is structured in a much more complex way than that of prokaryotes, even if the latter is far from being just a double helix of DNA suspended in the cell sap. Eukaryotic DNA is organized into nucleosomes--complexes of histones H2a, H2b, $H_3$, and $H_4$ wrapped around DNA. The chromatin fiber, with its nucleosomes and its acidic proteins, is folded, probably with the help of histone H1, into a chromatid. Before the chromatid is replicated, there is only one chromatid (unit chromatid) per chromosome, at least according to the widely accepted mononeme theory. This unit chromatid is thought to contain a single thread of the DNA double helix and to be continuous in each chromosome (at least in yeast and *Drosophila*). At mitotic metaphase, the chromosome is made up of two (sister) chromatids.

Chromatids are arranged into a series of chromomeres, which have the appearance of characteristic bands in polytene chromosomes, namely chromosomes formed by hundreds of chromatids originated by internal replication. Structurally, chromomeres correspond to local packings of chromatin; functionally, they have been classically considered as units of mutation, replication, and transcription, corresponding to Mendelian genes. These concepts are being further studied at the present time. In any case, chromomeres contain much more DNA than required by a gene coding for an average protein.

Another major difference between prokaryotic and eukaryotic cells is the segregation, in eukaryotes, of parts of the genome into cytoplasmic organelles (mitochondria, chloroplasts). This DNA seems to be structured according to the prokaryotic rather than the eukaryotic model.

## Genome Organizations

The prokaryotic genome is basically made up of a series of structural genes coding for messenger RNA, (and therefore polypeptide chains), ribosomal RNA, and transfer RNA. Promoters (sequences binding RNA polymerase) and operators (sequences binding regulatory proteins, such as repressors) are found at the beginning of each polycistronic transcription unit or operon. A prokaryotic genome has a single site for the initiation of replication, which proceeds bidirectionally; in other words, it is a replicon. Much of our knowledge of the organization of the prokaryotic genome comes from work on the lactose operon of E.coli. Recently, a whole genome, that of the bacteriophage ΦX 174, has been completely sequenced. These investigations have shown that overlapping genes exist in this genome, the same triplets being read in different frames.

The organization of the (nuclear) eukaryotic genome is far more complex. Its replication starts at multiple initiation sites. These multiple sites fulfill an absolute need since, without them, the replication of such a large genome would slow down cell division to an impossibly low rate. The major difference between prokaryotic and eukaryotic genomes, however, is in the organization of genes. Here, the important point is that the generally very large increase in genome size shown by eukaryotes as compared with bacteria is not accompanied by a proportionate increase in the number of different proteins encoded. For instance, in the early sea urchin embryo, only 37 million nucleotide pairs appear to be expressed as polysomal messenger RNA; this corresponds to 4% of the haploid genome, or to about 30,000 genes whose average size is equal to 1,200 base pairs. In adult sea urchin tissue, this number drops down to only 2,000 to 5,000 gene equivalents (6,14). This means that the sea urchin DNA expressed in the early embryo is only three times more than that expressed in E.coli, whereas that expressed in adult tissue corresponds to less than half the DNA expressed in E.coli.

These facts raise the problem of the role of the "excess DNA" of eukaryotes. Three general kinds of explanation have been proposed: 1) One possibility is that eukaryotic genes are "serially repeated" up to 1,000 times along the DNA duplex. The nucleotide sequence of each of the repeated or "slave" genes is brought into accord with that of a "master" gene at least once per generation, possibly during the meiotic prophase, by an unspecified rectification mechanism. Because of rectification, only mutational alterations in the master gene are ever detected. This possibility has been disproven as a general model for the organization of the eukaryotic genome on several grounds. It is true, however, that a number of "repeated genes" exist in eukaryotes: ribosomal genes, t-RNA genes, 5S RNA genes, and histone genes. Apart from these, all other genes which have been probed, (globin, ovalbumin, immunoglobulin, fibroin, etc.), have been found to be present in a single copy or a very small number of copies. 2) A second possibility is that the "excess DNA" is largely used for regulatory purposes; as we know, structural genes have not considerably increased in number during evolution, but it is conceivable that the amount of regulatory DNA has increased in amount. If such is the case, one might expect that such regulatory sequences are repeated a large number of times in the genome. 3) Finally, the "excess DNA" might be there for completely different purposes: for maintenance of chromosome structure, or as material to be used in evolution, meiosis, recombination, etc. This and the previous possibility are, of course, not mutually exclusive.

The c-value paradox, as the "excess DNA" problem has been called, along with the difficulties of eukaryotic genetics, justifies approaches at the molecular level, in which DNA sequences are directly investigated. Most of the information available so far is derived from the study of the kinetics of DNA renaturation.

## THE ORGANIZATION OF THE EUKARYOTIC GENOME AS SEEN BY THE KINETICS OF DNA RENATURATION

The reannealing of separated complementary single strands of DNA ideally follows a second order kinetics. For a given initial DNA concentration, and a certain fragment size, the half-time of reassociation should be proportional to the number of different types of fragments present and thus to the genome size (26,3). This expectation is exactly borne out in the case of the viral and bacterial genomes, which are characterized by a unique copy DNA sequence. Eukaryotic DNAs show complex renaturation kinetics and can usually be resolved into fast, intermediate, and slow-renaturing components (Figure 1). Renaturation kinetics studies, very largely based on a technique developed in our laboratory (13), have contributed two major points to our understanding of the organization of the eukaryotic genome: 1) A large fraction of the genome, usually 50-70%, consists of unique or single-copy sequences. These sequences show a very slow renaturation kinetics. The rest of the DNA renatures at faster rates, since it is made up of repetitive sequences with degrees of reiteration ranging from 10 to 1,000 copies to over 100,000 copies. The latter have been shown to correspond to "satellite" DNAs, which can usually be separated by density gradient centrifugation (see below). Some very fast-renaturing material, following first order

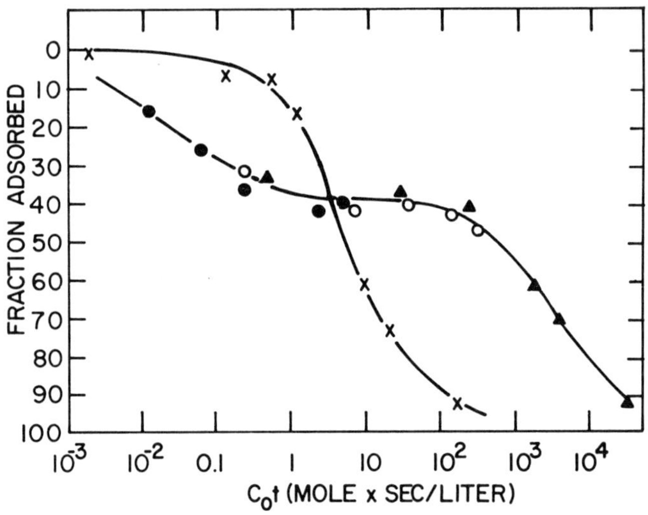


Figure 1. The kinetics of reassociation of calf thymus DNA with hydroxyapatite. Crosses are radioactively labeled E. coli DNA (3).

kinetics, has also been shown to exist in the eukaryotic genome. These fragments can fold back on themselves since they contain palindromic nucleotide sequences; they usually represent a few percent of eukaryotic DNA. 2) The relative arrangement of repetitive and nonrepetitive sequences can be investigated by incubating labeled DNA, sheared to various fragment lengths, with excess 450-nucleotide fragments of unlabeled DNA at a low Cot value (Cot equals the product of initial DNA concentration multiplied by reaction time), allowing reassociation to proceed, and following the binding of labeled DNA to hydroxyapatite (Figure 2). Analysis of data for *Xenopus* DNA (7) showed that a little more than 50% of the genome consists of closely interspersed repetitive and nonrepetitive sequences. The average length of the repetitive sequence elements is 300 ± 100 nucleotides, while the nonrepetitive sequences separating adjacent repetitive sequence elements average 800 ± 200 nucleotides. The remainder of the DNA is mainly nonrepetitive, though most of it contains rare

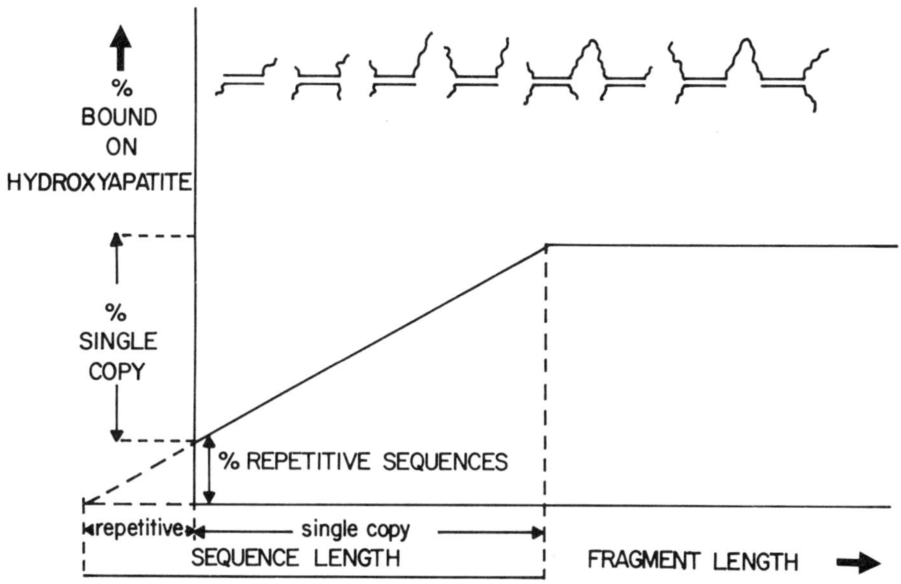


Figure 2. Schematic representation of the effect of fragment size on the fraction of DNA containing repetitive sequence elements, where interspersion is present. The curve results from a model calculation for a uniform length of repetitive sequences interspersed with a uniform length of nonrepetitive sequences (single copy sequences).

Unpaired regions; = paired; paired regions may be terminated by shear or end of repetitive regions (modified from reference 15).

interspersed repetitive elements spaced at a minimum of 4,000 nucleotides apart. A high degree of order in the arrangement of DNA sequences in the Xenopus genome is suggested by the alteration of relatively brief repetitive sequences and longer nonrepetitive sequences; while the latter have the size of single structural genes, the former could serve as sites for the specific binding of regulatory macromolecules. Evidence has been presented that most structural genes expressed in the gastrula of the

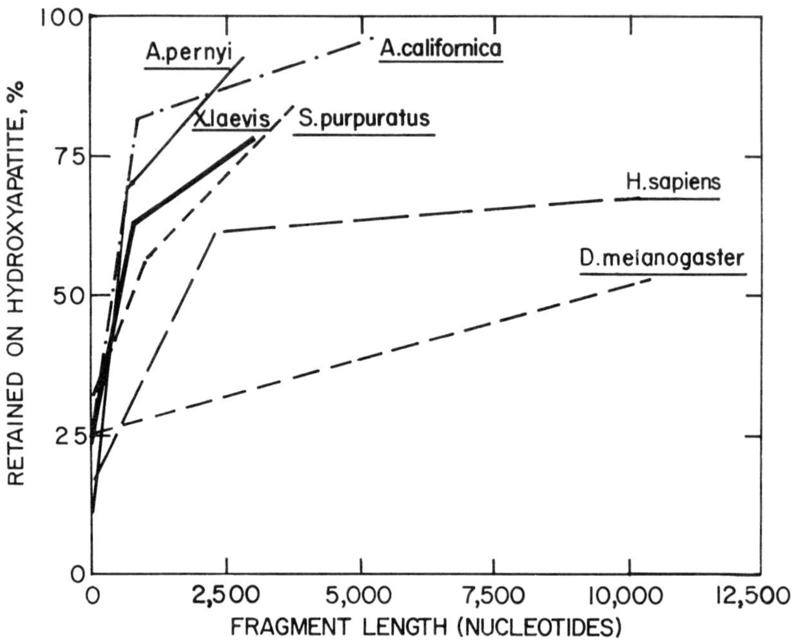


Figure 3. Interspersion data from several eukaryotic genomes (from references 1,7,10,15,19,22).

sea urchin (an organism having the same pattern of genome organization as *Xenopus*; see Figure 3) are adjacent to repetitive DNA sequences. This is suggested by the ability of single copy sequences adjacent to repetitive sequences to hybridize with labeled gastrular message preparations (8). Interestingly, nuclear RNA from sea urchin gastrulae has a single-copy sequence complexity ten times that expressed in polysomal message and is a transcript of interspersed repetitive and nonrepetitive sequences (23,24). Its function is unknown, but possibly it is of importance to the system of transcription regulation (16).

About 20 species, widely separated phylogenetically and including both deuterostomes and protostomes, show the *Xenopus* pattern (6,10,14). (Figure 3 and references 1,21). Among insects, one dipteran (*Musca domestica*) and a lepidopteran (*Antherea pernyi*) show the *Xenopus* pattern, while another dipteran (*Drosophila melanogaster*) and a hymenopteran (*Apis mellifera*) show a quite different pattern.

There are two major problems with these results. The first is the completely different genome organization exhibited by *Drosophila* and *Apis* compared with all other species investigated, which include *Musca*, a dipteran like *Drosophila*. The second problem is that the interspersion pattern of eukaryotes as widely separated as protostomes and man appears to be similar, as judged by the kinetics of DNA renatu ration. There is no explanation at present for the first problem. In contrast, the second problem seems to be mainly due to the limitations of the technique, particularly as used on unfractionated eukaryotic genomes.

## THE ORGANIZATION OF THE EUKARYOTIC GENOME AS SEEN BY THE ANALYSIS OF DNA IN DENSITY GRADIENTS

It has been well known since the classical work of Meselson, Stahl, and Vinograd (4) that, when centrifuged to equilibrium in CsCl density gradients, bacterial DNAs show unimodal, symmetrical bands, whereas calf thymus DNA exhibits a multimodal, broad, asymmetrical band. Bacterial DNAs, which have essentially statistical nucleotide sequences, show buoyant densities in CsCl density gradients which are linearly related to their base composition (25). This was thought to suggest that the CsCl band profile of calf thymus DNA is due to molecules differing in base composition. It was later realized, however, that the buoyant density in CsCl depends, in fact, not upon base composition, but upon short nucleotide sequences (3,5). Density gradient centrifugation provides, therefore, the basis for a very effective fractionation procedure of complex DNAs like eukaryotic DNAs, particularly when the procedure is performed in the presence of sequence-dependent ligands, like $Ag^+$, actinomycin, and BAMD [3,6-bis (acetatomercurimethyl) dioxane].

For the reasons just given, density gradient centrifugation methods have led to the separation of 1) "satellite" DNAs, formed by short, repetitive sequences, also called simple-sequence DNAs;

and 2) repetitive genes which contain short, repetitive sequences in the "spacers" separating the transcribed units (ribosomal genes, t-RNA genes, 5S RNA genes, and histone genes).

Both series of DNA components, the simple-sequence DNAs and the repetitive genes, are the best known eukaryotic DNA components. They can be easily prepared and their study by a number of techniques has made big advances in recent years. While the biological role of repetitive genes is well understood, that of simple-sequence DNA components is not. It is known, however, that they are not transcribed and that they are generally localized near the centromeres of chromosomes. It has been suggested that the simple-sequence DNAs play a structural role in meiotic pairing, in chromosome organization, and in recombination.

Both "satellite" DNAs and repetitive genes generally represent just a few percent of total eukaryotic DNA. Therefore, it is evident that all the unsolved problems reside in the so-called "main-band" DNA. Using centrifugation of DNA in $Cs_2SO_4$-$Ag^+$ density gradients, we not only found that we could separate four "satellite" components and several minor components, but we also discovered four years ago (9) that the main band of calf DNA, (molecular weight 13 million), was not formed by genome fragments having a continuous distribution of their average G+C-content, as was commonly believed, but by three distinct families of fragments. These were characterized by buoyant densities in CsCl equal to 1.697, 1.704, and 1.709 $g/cm^3$, and represented about 50, 25, and 10% of the DNA respectively (Figure 4). These three components gave rise to symmetrical CsCl bands, which exhibited standard deviations close to those of bacterial DNAs of comparable molecular weight; these components were mainly responsible for the asymmetry of the CsCl main band of calf DNA.

Three important questions were raised by the presence of these major DNA components in the calf genome: their evolutionary origin, their relative arrangement and their intramolecular heterogeneity. A detailed analysis of DNA preparations from 25

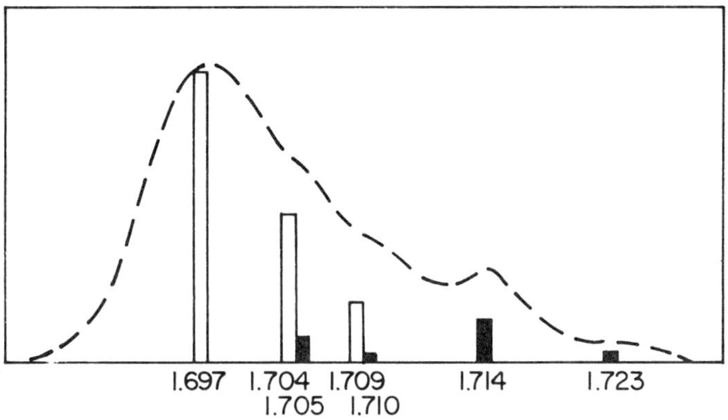


Figure 4. A histogram of the DNA components of the bovine genome. The height of the bars is proportional to the percentage of each component, filled bars correspond to the sharp melting; empty bars to the broad melting components. The broken line is an enlarged band pattern in CsCl of total calf DNA (1).

eukaryotes (23), ranging from yeast to man, led us to the recognition of phylogenetic differences at the macromolecular level in the organization of eukaryotic genomes. The differences concerning the main band DNA are: 1) the three major components, first observed in calf DNA, were also found in the other ten mammalian DNAs investigated; similar components appear to be present in avian genomes; 2) the DNAs of reptiles, amphibians, and fish showed a much lower and decreasing asymmetry of their CsCl bands compared to both mammalian and avian DNAs; 3) essentially symmetrical bands in CsCl were exhibited by three invertebrate DNAs; 4) the DNAs from three unicellular eukaryotes exhibited perfectly symmetrical bands in CsCl and could not be resolved into discrete components (Figures 5,6).

A novel feature of the main component of some mammalian DNAs is that it is made up of two subcomponents, which significantly differ in both buoyant densities and relative amounts in species which are

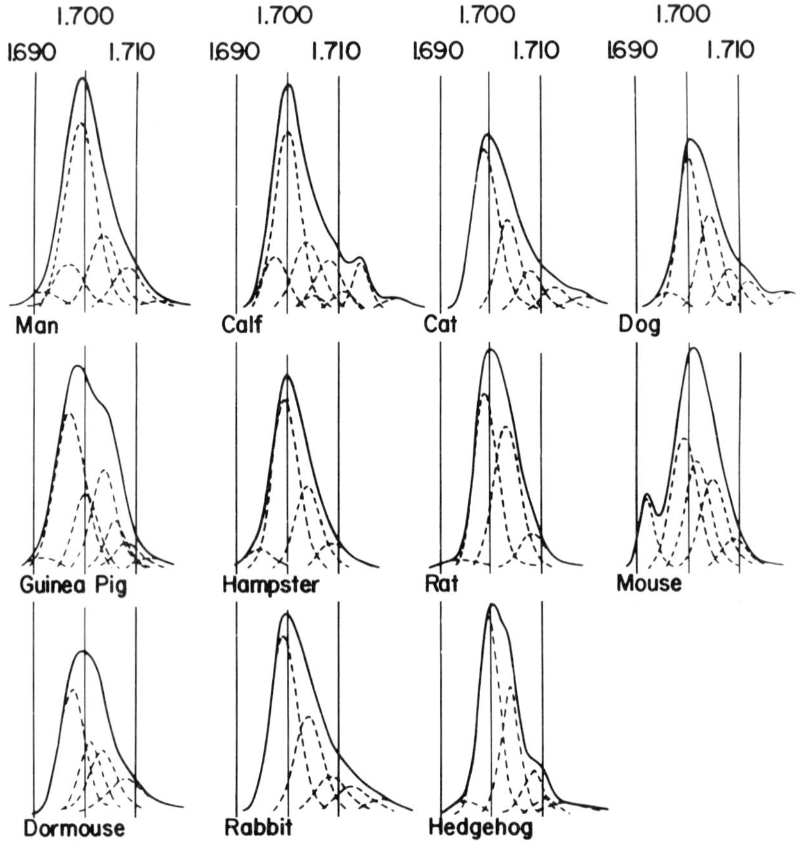


Figure 5. CsCl analysis of DNAs from 11 mammals (24).

evolutionarily not far from each other, and, more strikingly, in different tissues from the same species. Such differences are not due to differences in breed or sex since they have been seen in tissues from a single calf fetus. Two important open problems concerning the 1.704 and 1.709 $g/cm^3$ components seen in mammals and birds are: first, whether they are homologous in the two orders; and, second, what their biological significance is, e.g., whether they are associated with new functions like homeothermy or amnios formation.

The other two questions were answered by studying (19) a mammalian DNA in the molecular weight range of $1 \times 10^6$ to more than $200 \times 10^6$.

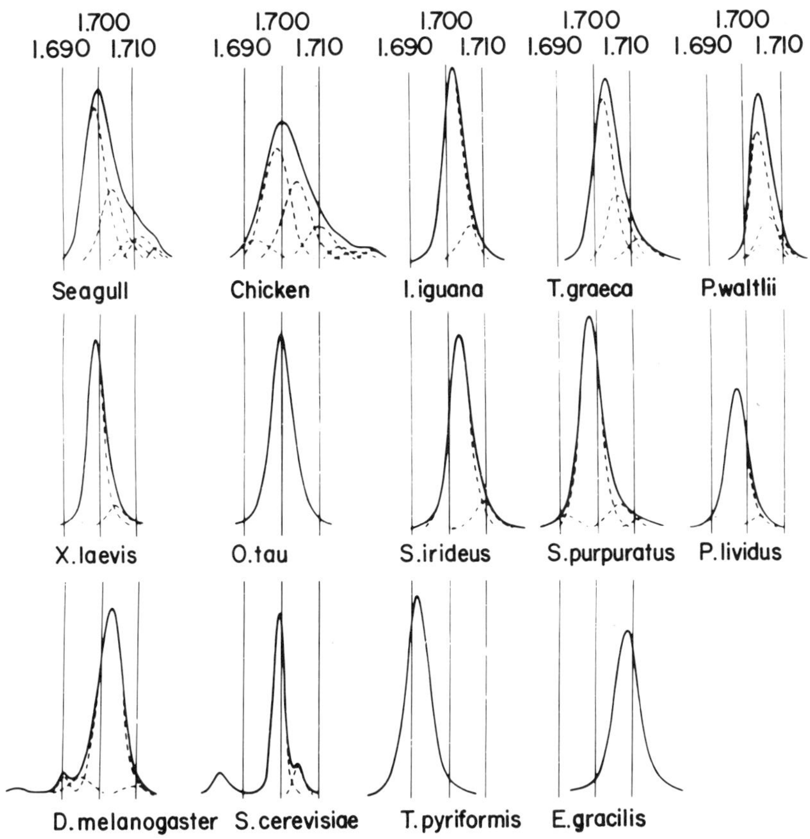


Figure 6. CsCl analysis of DNAs from birds, reptiles, amphibia, fishes, invertebrates, and unicellular eukaryotes (24).

The relative amounts and the buoyant densities of the major components did not vary in this size range indicating that the components are not intermingled with each other over very great lengths of DNA and that their internal sequence heterogeneity is not yet apparent at rather low molecular weight. In contrast, the major components of degraded amphibian DNAs showed a very different pattern compared with the undegraded DNAs, particularly in the case of amphibia having a very large genome size; changes in buoyant densities and amounts of some components, the disappearance of some components, and the appearance of new components were observed. Finally, the degraded *Drosophila* and *Saccharomyces* DNAs were similar to the undegraded preparations in that no components could be resolved within the main bands.

In conclusion, density gradient fractionation of native eukaryotic DNAs is an extremely powerful technique which probably provides the best first step to study the organization of eukaryotic genomes. Its possibilities are far from being fully explored, as shown by our recent work on the calf genome (2,20) combining $Cs_2SO_4$-BAMD and $Cs_2SO_4$-$Ag^+$ density gradient centrifugation, which has led to the preparation of seven simple-sequence and ten minor components. Obviously, this approach can be combined with a number of other techniques like renaturation kinetics studies, restriction enzyme degradation, cloning of DNA fragments, and hybridization experiments.

## REFERENCES

1. Angerer, R.C., Davidson, E.H., and Britten, R.J.: Cell, 6: 29, 1975.

2. Bernardi, G.: Nature, 206: 779, 1965.

3. Britten, R.J., and Kohne, D.E.: Science, 161: 529, 1968.

4. Corneo, G., Ginelli, E., Soave, C., and Bernardi, G.: Biochemistry, 7: 4373, 1968.

5. Cortadas, J., Macaya, G., and Bernardi, G.: Eur J Biochem (in press).

6. Crain, W.E., Davidson, E.H., and Britten, R.J.: Chromosoma, 59: 1, 1976.

7. Davidson, E.H., Hough, B.R., Amerson, C.S., and Britten, R.J.: J Mol Biol, 77: 1, 1973.

8. Davidson, E.H., Hough, B.R., Klein, W.H., and Britten, R.J.: Cell, 4: 217, 1975.

9. Davidson, E.H., Klein, W.H., and Britten, R.J.: Dev Biol, 54 (in press).

10. Efstratiadis, A., Crain, W.R., Britten, R.J., Davidson, E.H., and Kafatos, F.C.: Proc Nat Acad Sci USA, 73: 2289, 1976.

11. Filipski, J., Thiery, J.P., and Bernardi, G.: J Mol Biol, 80: 177, 1973.

12. Galau, G.A., Britten, R.J., and Davidson, E.H.: Cell, 2: 9, 1974.

13. Galau, G.A., Klein, W.H., Davis, M.M., Wold, B.J., Britten, R.J., and Davidson, E.H.: Cell, 7: 487, 1976.

14. Goldberg, R.B., Crain, W.R., Ruderman, J.V., Moore, G.P., Barnett, T.R., Higgins, R.C., Gelfand, R.A., Galau, G.A., Britten, R.J., and Davidson, E.H.: Chromosoma, 51: 225, 1975.

15. Graham, D.E., Neufeld, B.R., Davidson, E.H., and Britten, R.J.: Cell, 1: 127, 1974.

16. Hough, B.R., Smith, M.J., Britten, R.J., and Davidson, E.H.: Cell, 5: 291, 1975.

17. Macaya, G., Thiery, J.P., and Bernardi, G.: J Mol Biol, 108: 237, 1976.

18. Macaya, G., Cortadas, J., and Bernardi, G.: Eur J Biochem (submitted).

19. Manning, J.E., Schmid, C.W., and Davidson, N.: Cell, 4: 141, 1975.

20. Meselson, M., Stahl, F.W., and Vinograd, J.: Proc Nat Acad Sci USA, 43: 582, 1957.

21. Schildkraut, C.L., Marmur, J., and Doty, P.: J Mol Biol, 4: 430, 1962.

22. Schmid, C.W., and Deininger, P.L.: Cell, 6: 345, 1975.

23. Smith, M.J., Hough, B.R., Chamberlin, M.E., Britten, R.J., and Davidson, E.H.: J Mol Biol, 85: 103, 1974.

24. Thiery, J.P., Macaya, G., and Bernardi, G.: J Mol Biol, 108: 219, 1976.

25. Wells, R.D., and Blair, J.E.:
*J Mol Biol, 27: 273, 1967*.

26. Wetmur, J.G., and Davidson, N.:
*J Mol Biol, 31: 349, 1968*.

# *Possibilities of Cell Biology for Genetic Engineering*

Nils R. Ringertz

The term genetic engineering is now used to describe experiments designed to modify the genetic material of bacteria, viruses, and other prokaryotic microorganisms. This technology owes its origin to recent progress in nucleic acid biochemistry and molecular biology. Thus, the discovery of restriction enzymes has made it possible to prepare DNA fragments in a form which makes it easy to recombine and join fragments from two different organisms into recombinant DNA molecules. Using these methods it is possible to construct specific gene sequences which can then be injected into bacteria where they will be replicated and thus amplified as the bacteria multiply. The consequences this may have for the biological properties of bacteria is presently the subject of intense study and debate. In view of the rapid progress in genetic engineering with microorganisms it is now of interest to consider to what extent such manipulations can be performed with animal cells. It is also necessary to discuss the potential benefits and risks that such experiments may entail.

Recent advances in the field of somatic cell genetics have resulted in techniques which make it possible to generate animal and plant cells with altered genomes (Figure 1) such as mutant cells with specific genetic defects, virus-transformed cells in which viral genes have been added to the cellular genomes, hybrid cells containing genetic material from two different cell types, and reconstructed cells in which the nucleus may be derived from one cell type and the cytoplasm from another.

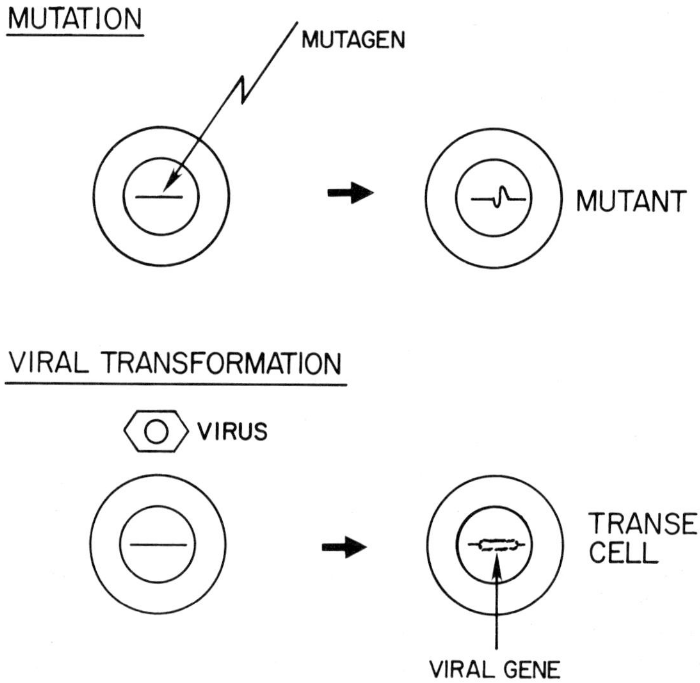


Figure 1. Schematic illustration of mutation and viral transformation.

The aim of this presentation is to discuss possibilities of preparing animal cells with altered genomes. For references and a more detailed discussion of this subject see (1).

## MUTANT CELLS

The term mutant is used to describe cells or cell lines which show a stable and heritable alteration in phenotype. Mutants can be isolated by culturing cells from patients suffering from inherited metabolic disorders or by treating normal ("wild-type") cells *in vitro* with mutagens. In the latter case, exposure to mutagens (X-rays or chemical mutagens) is followed by selection for mutant cells. Often one uses special culture media or drugs in order to kill normal cells and allow a special class of mutants to accumulate. Drug-resistant, enzyme-deficient mutants have been used in testing the relative importance of different pathways of purine and pyrimidine synthesis and in studies of metabolic cooperation and cross-feeding between cells. *In vitro* studies of cells sensitive or resistant to cytostatic drugs have provided information about the mechanisms of action of these drugs. This type of information is important in cancer chemotherapy. Auxotrophic mutants, that is cells requiring special nutrients not required by normal cells, have been used for genetic analysis of nutritional requirements. Temperature-sensitive mutants are cells which grow at certain temperatures (permissive temperatures) but not at other temperatures (nonpermissive temperatures) at which normal cells multiply. Many of these mutants appear to have slight DNA base changes which cause single amino acid substitutions in proteins, thereby causing the proteins to become nonfunctional when the temperature is changed from the permissive to the nonpermissive range. The protein affected may be one which is necessary for DNA, RNA, or protein synthesis in general or one which has a specific enzymatic function in a metabolic pathway. Other temperature sensitive mutations affect complex cell functions, for instance, progress through the cell cycle, mitosis, susceptibility to viral infection, or ability to undergo transformation by tumor viruses. The study of mutant cells by biochemical techniques makes it possible to test the direct involvement of a specific gene product in the control of a given phenotype. Furthermore, mutant cells may be used in cell fusion experiments to facilitate the isolation of hybrid cells. As will be discussed below this represents an important genetic technique which makes it possible to explore the organization and function of genes in human and animal cells.

## VIRUS TRANSFORMED CELLS

Treatment of normal cells *in vitro* with DNA or RNA tumor viruses can cause *transformation*, a heritable alteration in growth regulation. This process is believed analogous to *in vivo* transformation of normal cells to tumor cells. Cells which have been transformed continue to multiply *in vitro* after normal cells have undergone senescence and have stopped multiplying. Usually transformed cells do not contain detectable quantities of infectious virus. Using nucleic acid annealing, it is, however, possible to detect a small number of viral genomes which are covalently linked to host cell DNA (Figure 1). In some transformed cell lines the viral genomes are defective and correspond only to one or two genes. One of these genes is believed to specify the transformed phenotype and a special virus-induced nuclear antigen. Further studies of viral transformation genes will no doubt provide valuable information about virus-induced carcinogenesis.

## CELL HYBRIDS

The technique of cell hybridization involves a spontaneous or induced fusion of two different types of cells (A and B) into multinucleate heterokaryons containing A and B type nuclei within a common cytoplasm (Figure 2). Most heterokaryons die soon after fusion but some of those which contain only one A nucleus and one B nucleus survive and complete a normal cell division. The binucleate heterokaryons then give rise to two mononucleate hybrid cells (synkaryons), each of which contains a complete set of A and B chromosomes. Many types of hybrids show great vitality and are capable of multiplication in tissue culture for many years.

A wide variety of animal, human, and even plant cell types, have been used as parental cells in these fusions. When cells of different organisms are fused interspecific hybrids are produced.

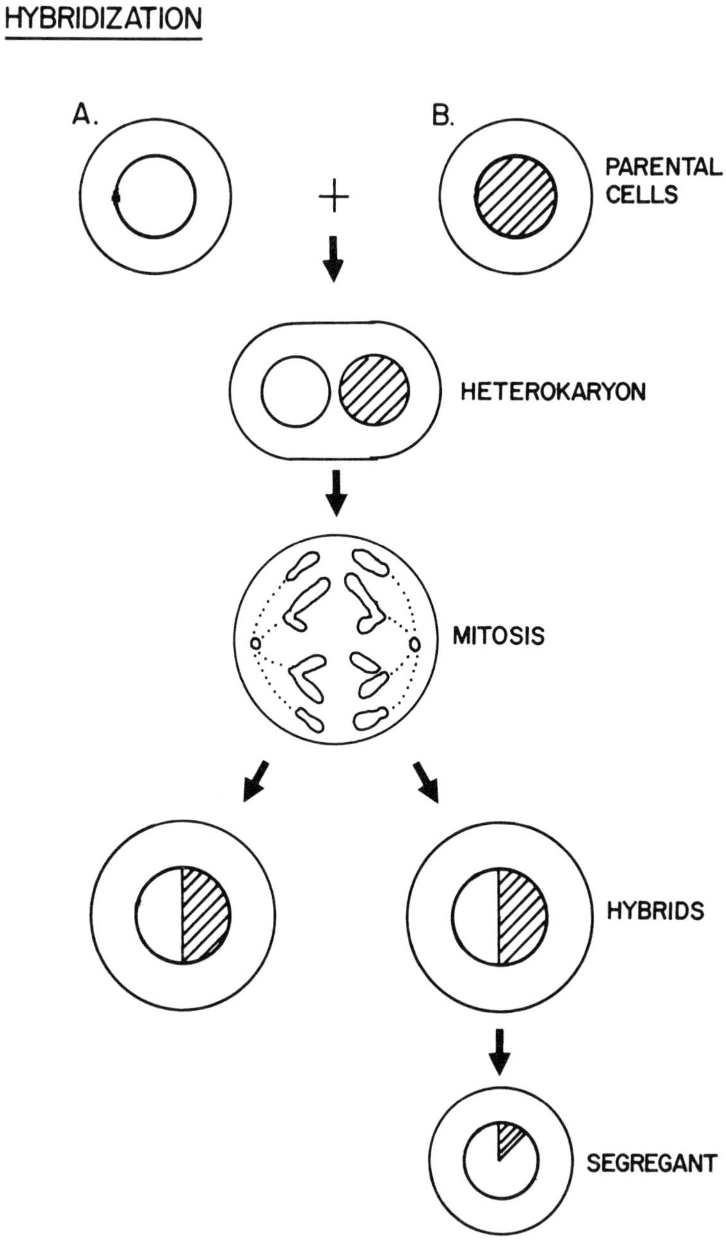


Figure 2. Schematic summary of somatic cell hybridization.

> Fusion of two different cells, A and B, results in a binucleate heterokaryon which in some cases divides to form two mononucleate hybrids (synkaryons) each of which contains a complete set of A and B chromosomes. If such hybrids are cultured over longer periods of time there may be elimination of chromosomes from one of the parental cells (chromosome segregation). A segregant may lose practically all chromosomes from the B cell while at the same time retaining a complete set of A chromosomes.

In these cases, the parental cells differ at least with respect to genotype if not phenotype as well. Intraspecific hybrids are obtained by fusing two different cell types from the one species. In these cases the parental cells share a common genotype but differ in morphological, immunological, or functional properties, that is, in phenotype.

The most common method of inducing cell fusion is to add inactivated Sendai virus to a suspension or monolayer of cells. Several chemical methods of inducing cell fusion have been suggested. These involve the use of lysolecithin, polyethyleneglycol, and artificially produced liposomes.

Heterokaryons, the immediate products when two different cells fuse, have been useful in the analysis of nucleocytoplasmic interactions and the role of the cytoplasm in the control of nuclear activity. Fusion of cells differing in their activity with respect to DNA and RNA synthesis has shown that the active partner almost always stimulates the nucleus of the inactive partner to synthesize more RNA and/or DNA. There is suggestive evidence that the signals which trigger the increased activity are proteins which migrate from the cytoplasm into the inactive nucleus. The signals appear not to be species specific since active cells of mouse origin can stimulate inactive human, rabbit, or chick as well as mouse nuclei.

If one of the parental cells in a fusion happens to be in mitosis, while another is in interphase, the nucleus of the latter cell is induced to undergo premature chromosome condensation (PCC). The morphological appearance of PCC filaments varies depending on whether the interphase cell happens to be in the $G_1$, S, or $G_2$ phase. Therefore, the PCC phenomenon can be used to analyze the chromosome condensation cycle during interphase and factors which trigger mitosis. Furthermore, it represents a method of visualizing chromosomes in nondividing cells (e.g., sperm), and of scoring the immediate effects of chromosome-breaking agents.

Heterokaryons have been useful in the analysis of factors which regulate phenotypic expression and in gene complementation analysis of human genetic disease. Restoration of a missing function in a heterokaryon formed by the fusion of two mutant cells is an indication that the two cell types suffer from different genetic defects. Using this approach it has, for instance, been possible to show that the *Xeroderma pigmentosum* syndrome can be divided into four to six complementation groups. Each of these groups probably corresponds to a specific genetic lesion, representing an enzyme or factor required for the repair of UV-induced DNA damage.

In order to study mononucleate hybrid cells (synkaryons) it is usually necessary to separate them from the parental cells and also from any polyploid cells which may arise from the fusion of like cells. This can be achieved by single cell cloning or by using culture conditions which favor the growth of hybrid cells while inhibiting the parental cells. Selection for hybrid cells is commonly based on the use of mutant parental cells. The "HAT" selection system can be used as an example of this. Mutant cells deficient in hypoxanthine-guanine phosphoribosyl transferase ($HGPRT^-$) are obtained by selection with thioguanine or from patients suffering from the Lesch Nyhan syndrome. These cells are then fused with cells deficient in thymidine kinase ($TK^-$).

The latter type of mutant is obtained by selection with bromodeoxyuridine. The genetic defects of the mutant cells are of little importance during growth on normal tissue culture media, since the relevant enzymes are only involved in salvage (reserve) pathways for nucleotide synthesis. When the main biosynthetic pathways for purine and pyrimidine nucleotides are blocked by the folic acid analogue aminopterin, normal cells ($HGPRT^+$ $TK^+$) can survive if supplied with exogenous hypoxanthine and thymidine whereas the mutant cells die because of their inability to synthesize nucleotides from hypoxanthine ($HGPRT^-$ cells) or from thymidine ($TK^-$ cells). Hybrid cells formed from fusion of $HGPRT^-$ with $TK^-$ cells can be isolated by selection in a medium containing Hypoxanthine, Aminopterin and Thymidine (HAT medium). Hybrids contain one chromosome set which is $HGPRT^-$ but $TK^+$ and one that is $HGPRT^+$ and $TK^-$. Therefore, they are able to produce HGPRT and TK enzyme and to utilize exogenous hypoxanthine and thymidine for nucleotide synthesis. Thus when combined in one cell the two parental genomes complement each other and make it possible for the hybrid cell to survive on HAT medium. The unfused parental cells, however, are killed by this medium.

In addition to the HAT medium there are a number of other selective systems in which the parental cells are temperature sensitive or auxotrophic mutants. When the two parental cells differ with respect to their gene mutations, the hybrids can be isolated because of the gene complementation phenomenon. This type of analysis also illustrates an important application of the cell hybridization method, namely the genetic analysis of metabolic pathways and cell function.

Intraspecific hybrids (e.g., mouse + mouse) usually show chromosome numbers which approximate or are a little below the expected sum if one assumes that one cell of one type and one cell of the other type have fused (1 + 1 hybrids). As the cells are maintained in culture over long periods of time the chromosome number undergoes a slight decrease. Interspecific hybrids (e.g. mouse + man) on the other hand usually show extensive chromosome elimination. Presumably chromosome segregation in the hybrid population is due to abnormal mitoses and overgrowth of variants with a reduced chromosome complement. In interspecific hybrids chromosome segregation usually involves the preferential elimination of chromosomes of one species while the chromosomes of the other species are selectively retained. Thus, human chromosomes are selectively eliminated in man + rodent hybrids. Since many homologous isozymes of man and mouse can be distinguished by differences in their electrophoretic mobility it is possible to determine whether there is coexpression of both genomes or whether a human enzyme activity has been lost because of loss of a specific human chromosome. By analyzing the expression of a variety of human markers in many different clones it is possible to establish that some markers always occur together and that they are always lost as a group (linkage analysis). In many cases the retention or loss of a marker can be correlated with a specific chromosome (synteny testing). Hybrid cells can also be used to establish the linear order in which genes occur on individual chromosomes, measuring distances between genes and relating these data to the structure of the chromosomes (chromosome mapping). One method of mapping genes is to induce massive chromosome fragmentation by treating one of the parental cells or the hybrids with chemical or physical mutagens, causing chromosome breaks. By using appropriate mouse mutants as one of the parental cells, and selective media on which the cells have to retain a specific human gene in order to survive, it is possible to obtain hybrid clones that retain a broken chromosome carrying the complementing human gene. Clones differ with respect to the size and break points of the chromosome fragment. Detailed cytological examination of the chromosome fragments and assaying for linked genes by isozyme analysis make it possible to map the exact localization of genes down to the molecular level.

One of the main applications of the cell hybridization technique in cell biology has been in the analysis of gene expression and cell differentiation. Hybrid cells with complete or reduced chromosome sets have been examined for general characteristics such as morphology, growth rate, contact inhibition and senescence, complex physiological and immunological properties, specific gene products such as enzymes, hormones, and immunoglobulins; and sensitivity to specific drugs. Any of these characteristics can be used as a marker if the two parental cells differ with respect to it. Obviously a distinction has to be made between properties common to all cells of a given organism (constitutive markers) and properties which are expressed only by certain differentiated cells (facultative markers). A number of different gene expression patterns have been observed: coexpression of constitutive and facultative markers, dominance or recessiveness of drug resistance markers, extinction of facultative markers and in some cases activation of new properties not expressed by the parental cells. Although these observations are important for our understanding of how gene expression and differentiation are controlled in eukaryotic cells they do not yet provide a clear picture of how gene activity is regulated in animal cells.

The technique of cell hybridization has also gained wide application in the analysis of malignancy and transformation. It is an important tool in tumor virology where it has been used to map virus integration sites, for virus rescue and virus detection, for analysis of factors determining the susceptibility of different cell types to viral infections, and for analysis of cellular mechanisms inhibiting viral replication or modifying viral gene expression.

## RECONSTRUCTION OF CELLS BY FUSION OF CELL FRAGMENTS

A new technique which promises to be of great interest in the study of gene regulation and nucleocytoplasmic interactions in animal cells involves fusion of cell fragments. The technique used for generating cell fragments is based on the use of the drug cytochalasin B. If animal cells are exposed to this drug, marked changes occur in the shape of the cells and frequently the nuclei are extruded into the tip of long protrusions which are connected to the main cytoplasm via narrow stalks. These stalks frequently break spontaneously, thus causing a loss of nuclei from some cells. The efficiency of enucleation can be increased to close to 100% by centrifugation. Usually glass discs with monolayers of cells are placed cell-side down in centrifuge tubes containing a cytochalasin solution. During centrifugation the nuclei are pulled out of the cytoplasms which remain attached to the discs. The cytoplasms (anucleate cells) have an abnormal shape immediately after centrifugation but if returned to normal culture medium they resume the form of the intact cell and continue to synthesize protein. The cytoplasms remain metabolically intact for 12 to 36 hours before they round up and die.

The nuclei, on the other hand, can be collected from the bottom of the centrifuge tubes. These nuclei differ from nuclei prepared by other techniques in that they are surrounded by a thin rim of cytoplasm and an intact plasma membrane. These cell fragments, which are referred to as minicells or karyoplasts, in many cases contain less than 10% of the cytoplasm of the intact cell and are capable of RNA and DNA synthesis for several hours after enucleation. Under the conditions of preparation and culture which we have used, minicells from several different cell types have been unable to regenerate a cytoplasm and to form cell colonies. Instead the minicells have undergone lysis and have died within 24 to 48 hours of enucleation.

Since nuclei and cytoplasms prepared by the cytochalasin enucleation technique are surrounded by an intact plasma membrane carrying receptors for Sendai virus they can be fused together with inactivated virus to give reconstituted cells (Figure 3). It is also possible to add cytoplasm

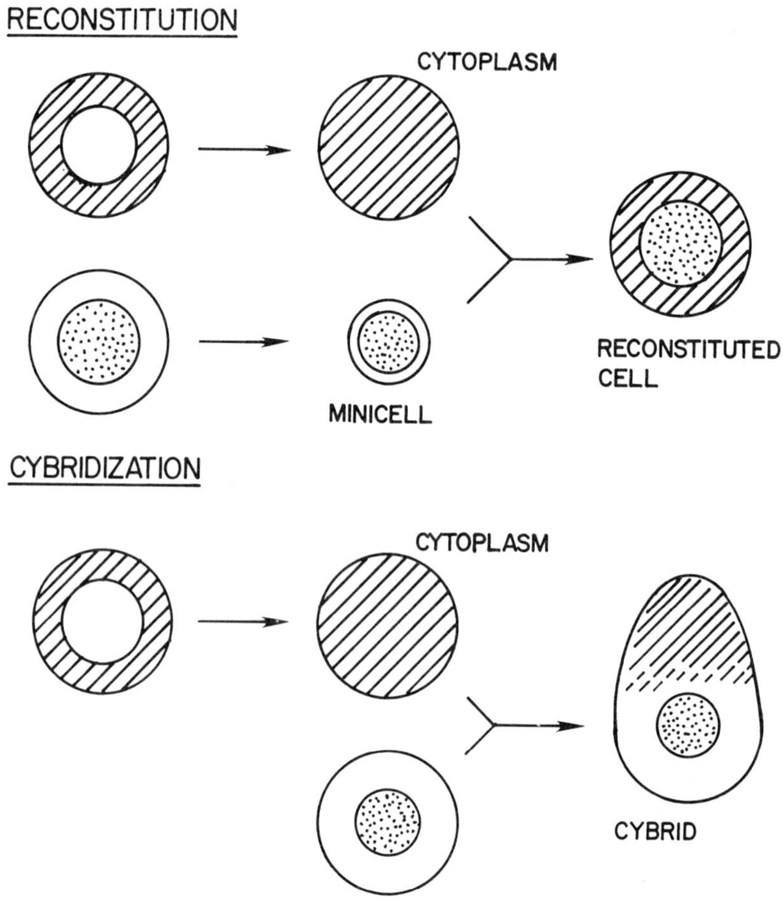


Figure 3. Enucleation of cells by cytochalasin treatment and centrifugation of animal cells gives two types of cell fragments: nuclei (minicells) and cytoplasms.

Both types of fragments are surrounded by an intact plasma membrane and may be fused to give a reconstituted cell. Viable cells may be reconstituted by combining nuclei and cytoplasms from different types of cells. If an enucleated cell (cytoplasm) is fused with an intact cell, the result is a cytoplasmic hybrid (cybrid).

from one cell to an intact cell. In this case the result is a cytoplasmic hybrid (cybrid), that is, a cell containing a cytoplasm which is a mixture of that of two different cells. One of the chief problems in such experiments lies in recognizing and distinguishing the intact parental cells, the reconstituted cells, and the cybrids. Four types of markers have been used to identify the origin of individual cells and small colonies arising from reconstructed cells: morphological markers (nuclear size and shape), artificially produced markers (ingestion of polystyrene beads of different size classes and/or labelled with different fluorochromes), DNA content, and functional markers. The functional markers (mutant cells defective in specific enzyme functions or resistant to specific drugs) are useful not only for identification purposes but also to obtain the progeny of the desired reconstructed cells by growth on selective media. Using this approach we have been able to reconstitute viable cells by combining nuclei and cytoplasms from different sublines of an established rat myoblast line. Also some combinations involving nuclei and cytoplasms from two different species have been successful. Thus viable cells capable of cell multiplication have been obtained by combining nuclei from rat myoblasts with cytoplasms from mouse fibroblasts. The fact that animal cells can be reconstructed by fusing two nonviable cell fragments from two different species and two different cell types indicates that it may be possible to generate a fairly wide spectrum of reconstituted cell types. Specific areas of research in which reconstitution techniques should prove useful include regulation of gene expression, the stability of the differentiated state, interactions between the nucleus and cytoplasm, and the dependence of mitochondria and other cytoplasmic organelles on nuclear genes.

Using the cytochalasin enucleation technique it is also possible to prepare another type of cell fragment. Microcells are prepared by centrifuging micronucleated cells in the presence of cytochalasin (Figure 4). Such cells can be

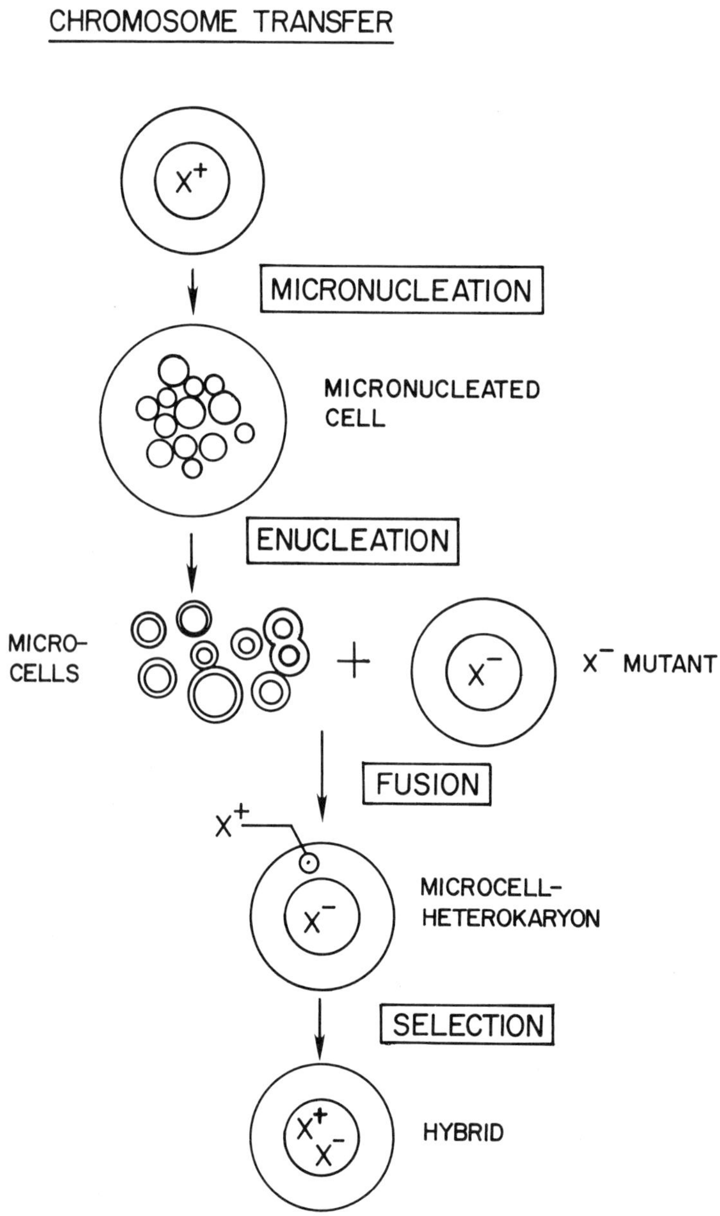


Figure 4. Transfer of chromosome(s) from one cell to another by microcell fusion.

Microcells are subdiploid miniature cells which are generated by first inducing micronucleation with microtubular poisons. Micronucleated cells are then enucleated to give microcells. The smallest microcells have a DNA content equivalent to single chromosomes. Fusion of microcells from normal cells with a mutant cell ($x^-$) can be used to introduce a complementing chromosome ($x^+$). By using appropriate selection methods one may be able to predetermine which microcell hybrids survive.

obtained in large numbers by treating normal mononucleated cells with microtubular poisons (colchicine, colcemid, vinblastine, etc.). After an extended metaphase, nuclear membranes reassemble around individual or small groups of chromosomes which are scattered throughout the cytoplasm of the dividing cell because of abnormal anaphase movements. As a result the genome is divided into a large number of micronuclei in which the chromosomes are dispersed into interphase chromatin. Enucleation of micronucleated cells results in subdiploid microcells containing one or several micronuclei surrounded by a rim of cytoplasm and a plasma membrane. The smallest members of a microcell population have a DNA content equivalent to single chromosomes, while the larger microcells clearly must contain more than one chromosome. As with minicells most of the subdiploid microcells are metabolically intact in the sense that they persist in culture for a short time and exclude trypan blue. Using inactivated Sendai virus, microcells may be fused with intact cells to give heterokaryons. There is tentative evidence that some microcell heterokaryons divide and give rise to mononucleate hybrid cells. Fusion with microcells, therefore, appears to offer a method by which a small part of the genome of one cell can be introduced into another cell. If the recipient cell is a mutant which is unable to grow on a selective medium it may be possible to isolate hybrids in which a complementing chromosome has been introduced by fusion with a microcell (Figure 4). Thus, although it will be difficult to fractionate microcells and obtain those containing a specific chromosome, and although fusion is likely to be random, it is possible to control which hybrids one obtains. Microcell hybrids containing mouse chromosomes and only one specific human chromosome could be useful in chromosome mapping. Other applications could be in gene complementation analysis designed to distinguish structural and regulatory mutations, and in the analysis of integrating sites for tumor viruses.

## CONCLUDING REMARKS

Recent progress in somatic cell genetics has made it possible to generate many different types of genetically altered cells. The techniques are based on the use of mutagens, viral transformation, cell fusion, and methods of preparing and recombining cell fragments. With these techniques it is now possible to analyze a large number of important biological and medical problems. Undoubtedly it is also possible to repair genetic defects in mutant cells by introducing complementary chromosomes from normal cells. In spite of this, it is not likely that such techniques will be useful for clinical medicine within the foreseeable future. The main reasons for this belief are a) that the problems involved in repopulating an individual suffering from a genetic disease with cells "repaired" _in vitro_ appear to be exceedingly difficult; there may also be a risk that the repaired cells will give rise to tumors; and b) there are at least 1,500 distinguishable human diseases which are known to be genetically determined though most of these diseases are rare or very rare. Furthermore, even if the patients show exactly the same symptoms they may differ with respect to their individual genetic lesions. The cost of mapping the nature of the genetic lesion in each individual and then designing a genetic therapy based on genetic engineering with cells or DNA would be prohibitive.

In view of the recent discussion about the dangers of recombinant DNA research it may be appropriate to ask if there are any dangers in genetic engineering with human or animal cells. So far there are no indications that there would be any serious risks in this type of research. On theoretical grounds it has been pointed out, however, that cell hybrids between animal and human cells may represent an intermediate host in which animal viruses can be adapted to grow on human cells by mutation or recombination with human cellular DNA. Clearly this possibility must be explored. It appears, however, to be a risk which may be controlled by using known methods for containing pathogenic microorganisms.

## REFERENCE

Ringertz, N.R., and Savage R.E.: Cell Hybrids. Academic Press, New York, 1976.

# Genetic Engineering: Achievements and Prospects

Charles Weissmann

Genetic engineering may be defined as the technology serving to generate viruses or organisms with a desired constitution in a directed fashion, rather than by trial and error. The goals to which the pursuit of genetic engineering may lead are: 1) elucidation of the relationship between structure and function of genetic material; 2) generation of microorganisms, plants, or animals with enhanced nutritional or economic value; or 3) the correction of certain genetic defects in humans.

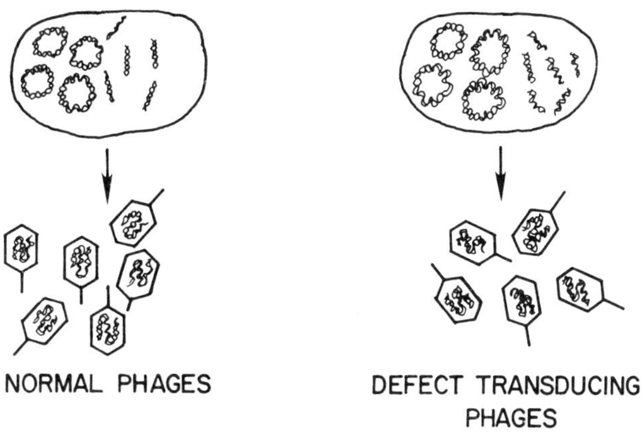


Figure 1. Normal and transducing phages.

## LEVELS OF BIOLOGICAL MANIPULATION

### Manipulations at the Level of the Intact Virus or Organism, Using Natural Sexual or Parasexual Mechanisms (Figure 1).

An example of this type of manipulation is the transduction of bacterial genes by phage $\gamma$. There exists a class of bacterial viruses, the temperate phages, which have the capacity of integrating their DNA into the chromosome of their host. As shown in Figure 1a, the penetrating viral genome is converted into a circular form and can then recombine with the circular genome of the host. This event generates an extended circular DNA which comprises viral and host genomes. In this so-called lysogenic state bacteria can multiply virtually indefinitely. Under certain circumstances (induction) the viral genome is excised from the host genome, at precisely the same points at which

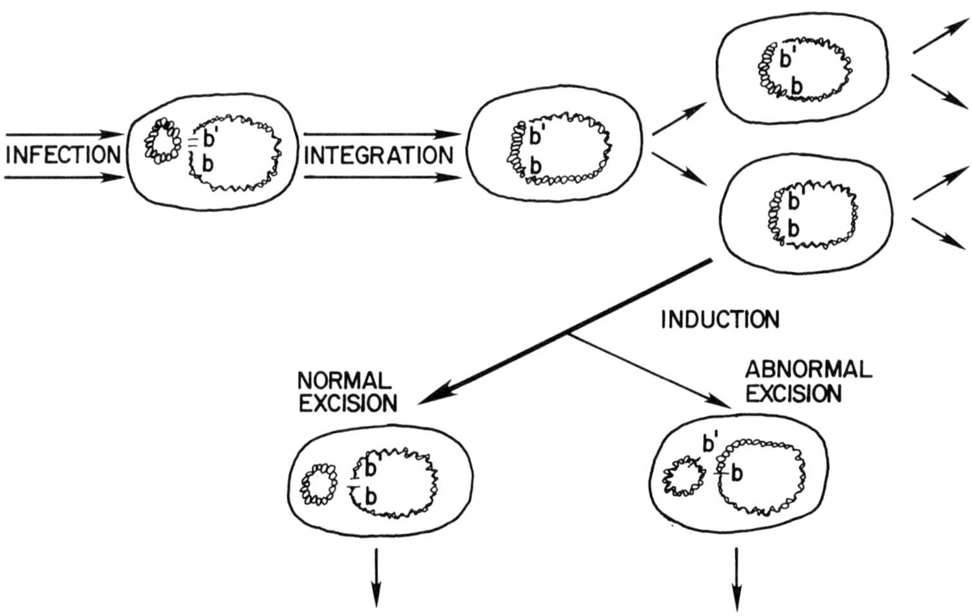


Figure 1a. Lysogenic infection.

it had been inserted, and the lytic cycle is initiated: the viral genome is multiplied, viral proteins are synthesized, and the two are combined to form new viral particles. However, occasionally an error occurs during the excision process and instead of the viral genome being neatly excised, part of the bacterial genome is removed, carrying with it bacterial genes located next to the insertion site, for example, genes from the galactose operon. In compensation some of the viral DNA is lost. Such viral particles can be enriched. If they are used to infect E. coli in which the galactose operon is inactive, the $\gamma$ gal can be inserted, along with the functional gal genes, rendering the bacterium competent to utilize galactose (Figure 1b). Genetic methods have been devised to transfer a variety of segments of E. coli DNA to such lysogenic phages or other carriers, such as F factors.

## Manipulations at the Level of Isolated Cells or Cell Organelles.

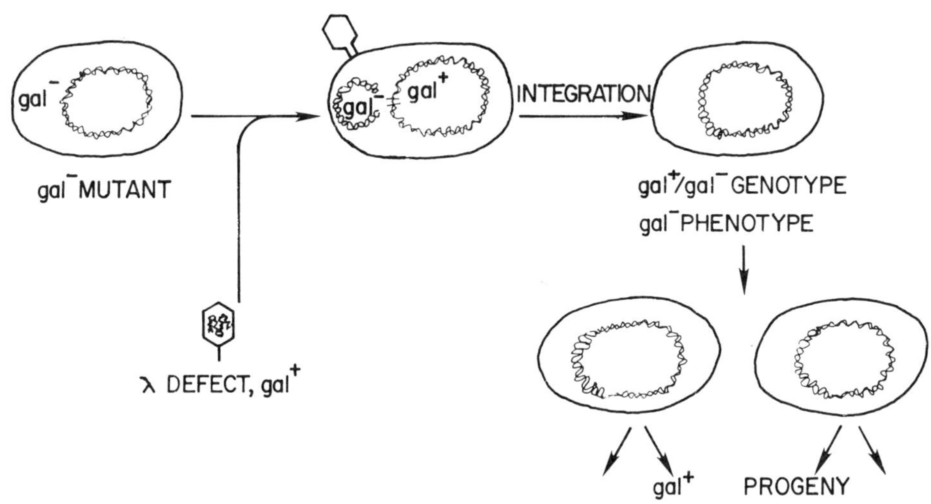


Figure 1b. Transduction.

Cell fusion technology (Figure 2). Cells isolated from intact organisms, usually grown in cell culture, are brought into close contact, whereupon the plasma membranes fuse into each other. This process can be greatly furthered by addition of UV-inactivated Sendai virus, or, more recently, polyethylene glycol, which catalyzes the interaction of the cell membranes. The resulting heterokaryon still possesses two nuclei, one from each parental cell, but when mitosis occurs, the chromosomes mix and give rise to a so-called synkaryon, a nucleus which comprises chromosomes from both parents. These synkaryons can be cultured further. On subsequent mitoses single chromosomes may be ejected, modifying the genetic constitution and consequently the phenotype of the cells. This approach has become extremely valuable for mapping genetic functions on eukaryotic, in particular human, chromosomes.

In the case of plant cells it is possible to fuse cells from different species and produce viable interspecies hybrids. Recently, it has even been possible to fuse plant cells with animal cells.

Transformation by Chromosomes. It has also proved possible to introduce single chromosomes into cells

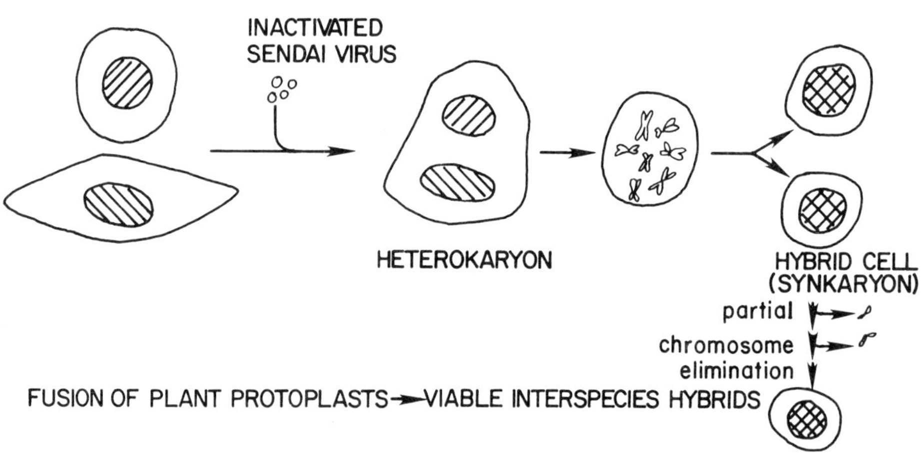


Figure 2. Cell fusion.

in a simple fashion, namely by exposing mutant cells incapable of synthesizing a particular enzyme, for example, hypoxanthineguanine phosphoribosyl transferase, to isolated chromosomes of normal cells. Cells which have taken up the foreign genetic material are now capable of synthesizing the missing enzyme, which can be characterized as being typical for the donor cell and not for the acceptor cell.

Haploid Cell Technology (Figure 3) . Haploid embryos can be grown from the anthers of plants in culture. From the haploid plants one can produce single cells, devoid of a cell wall (protoplasts). These can be mutagenized and cells with certain properties, such as resistance to toxins from plant parasites, can be selected. Since one is dealing with haploid cells, which have only one set of genes, any phenotypic property is immediately

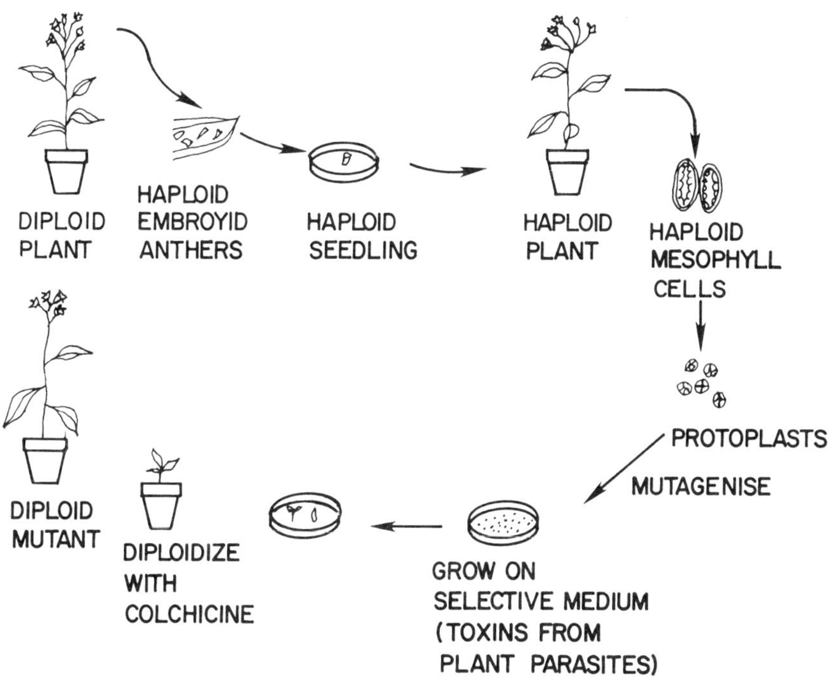


Figure 3. Haploid plant cell technology.

expressed, as in the case of bacteria and in contrast to the situation with diploid cells, where phenotypic expression is complicated by the existence of a second set of genes, which may mask expression of a mutated gene. Cells which have been selected for a certain property may then be converted into diploid cells by treatment with colchicine. These cells, which are homozygous for the mutated gene, can be grown into normal diploid plants. This powerful technology will undoubtedly allow important advances in the breeding of new, superior plants.

<u>Cloning of Vertebrate Organisms</u> (Figure 4). The nucleus of an oocyte was destroyed by UV-irradiation and the diploid nucleus of an intestinal cell of a tadpole was injected into the enucleated oocyte. With a reasonable frequency these oocytes developed into mature, fertile frogs. These experiments suggest possibilities for obtaining

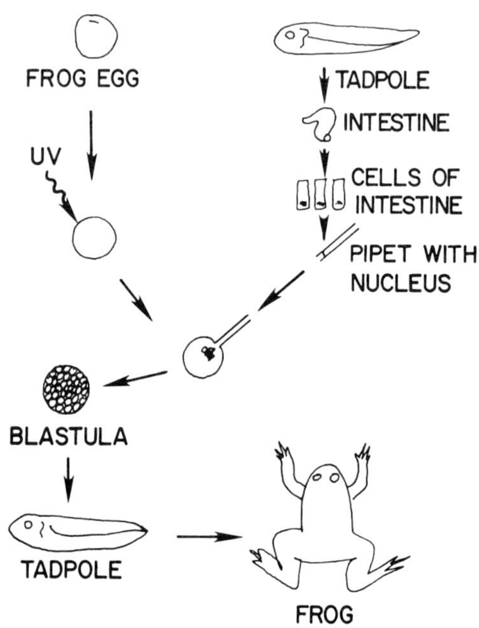


Figure 4. Nuclear transplantation.

populations of animals with genetic constitutions which have, so to speak, been pretested for their phenotype, an approach that could be valuable in animal husbandry. However, this approach has not yet been successful with mammals.

## Modifications of Genetic Material at the Molecular Level

Site-Directed Mutagenesis (Figures 5 and 6). This is a technique, developed in our laboratory, which allows the substitution of single nucleotides at predetermined positions of the genome. This method has been used to generate mutations in various biologically interesting positions of the genome of an RNA bacteriophage, Q . This phage has an RNA genome of about 4,500 nucelotides in length, which can be synthesized in vitro with a purified enzyme, Q replicase. Site-directed mutagenesis is carried out by incorporating a mutagenic nucleotide analog into a specific position of the strand complementary to the viral RNA. On replicating the minus strand a mutation

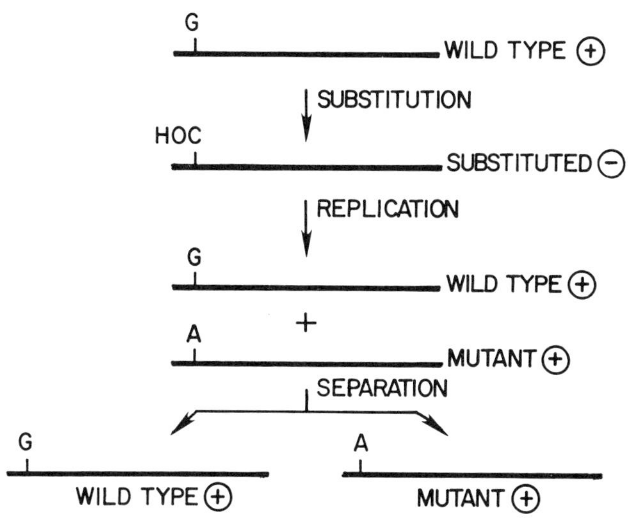


Figure 5. Site-specific mutagenisis.

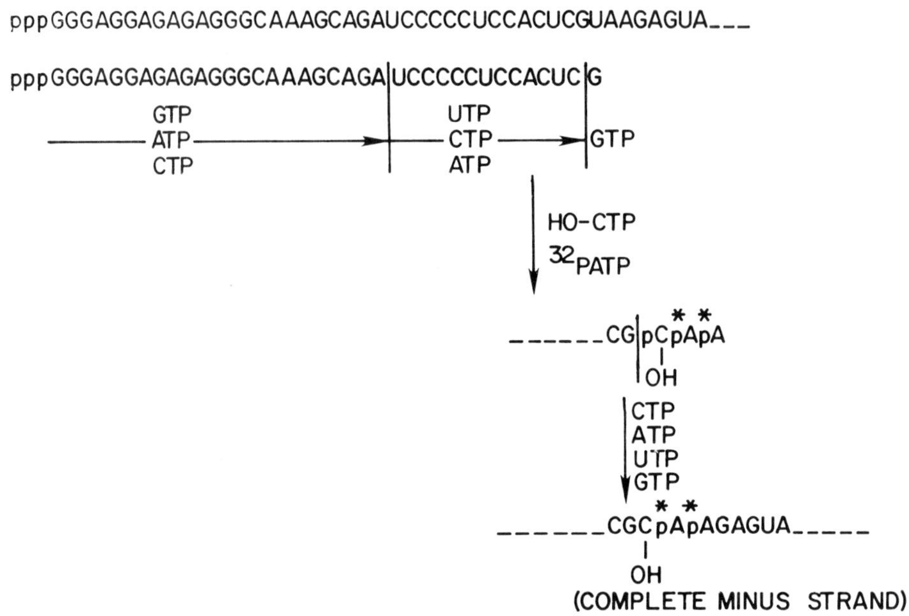


Figure 6. Substitution of HO-C for U in position 39 of the Q minus strand.

is generated in the position directed by the mutagenic analog (Figure 5). The insertion of the analog into the desired position is carried out by controlled enzymatic synthesis. One approach to this controlled synthesis is shown in Figure 6. Synthesis on the template is initiated by adding only three of the four required nucleoside triphosphates, namely GTP, ATP, and CTP, so that synthesis can proceed only up to the point where the missing substrate is required, namely position 23. The unused substrates are then removed and substituted by a second set of nucleotides, with UTP, which was previously missing, now present. This process of stepwise synthesis is repeated up to the point where the mutation is to be placed. Then the nucleotide analog is added, allowing the incorporation in the one desired position only, and the enzyme RNA complex is separated from the substrate. After that, the four normal ribonucleoside triphosphates are added and the complementary strands are completed. On using these substituted, purified complementary strands as template for the replicase, viral RNA with base substitutions in the desired positions was generated. Figure 7 shows two positions in the extracistronic region in which mutants have been generated. One of these gave rise to a viable mutant; the other was lethal. This general approach has also been used to generate mutations in the middle of the genome, at a ribosome binding site.

<u>Chimerization and Cloning of DNA (Figure 10)</u>. This is one of the most important recent advances in genetic engineering. It is based on two important techniques: a) the use of restriction enzymes, i.e., nucleases with a high specificity for

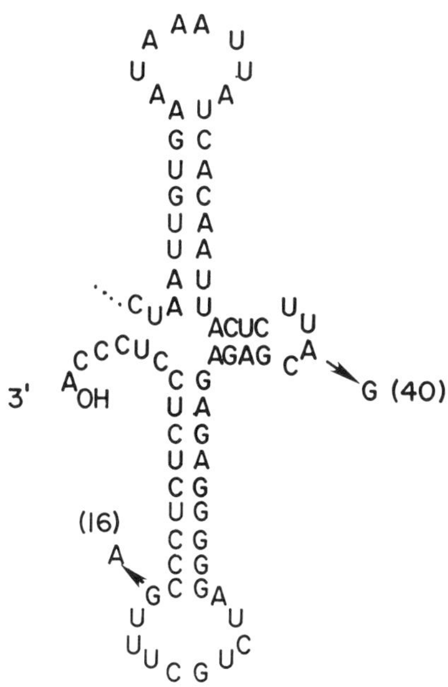

Figure 7. Generation of mutants in the extracistronic region.

certain nucleotide sequences, to cleave DNA at precisely defined positions and the use of a series of synthetic enzymes, which can elongate, repair, and link DNA fragments; and b) the use of plasmid or bacteriophage DNA as vehicles for the replication of specific DNA fragments.

Figure 8 illustrates the way a typical class II restriction enzyme, namely $EcoR_1$, works. It recognizes the specific sequence indicated and cleaves it in a staggered fashion, at the positions shown by the arrows. Since, for statistical reasons, the sequence recognized by the enzyme occurs only rarely in a DNA, the DNA pieces resulting from the enzymatic digestion are long, on the average around 6,000 base pairs, and may thus comprise one or more genes. The staggered cut introduced by the enzyme yields so-called cohesive ends, i.e., ends that can overlap by

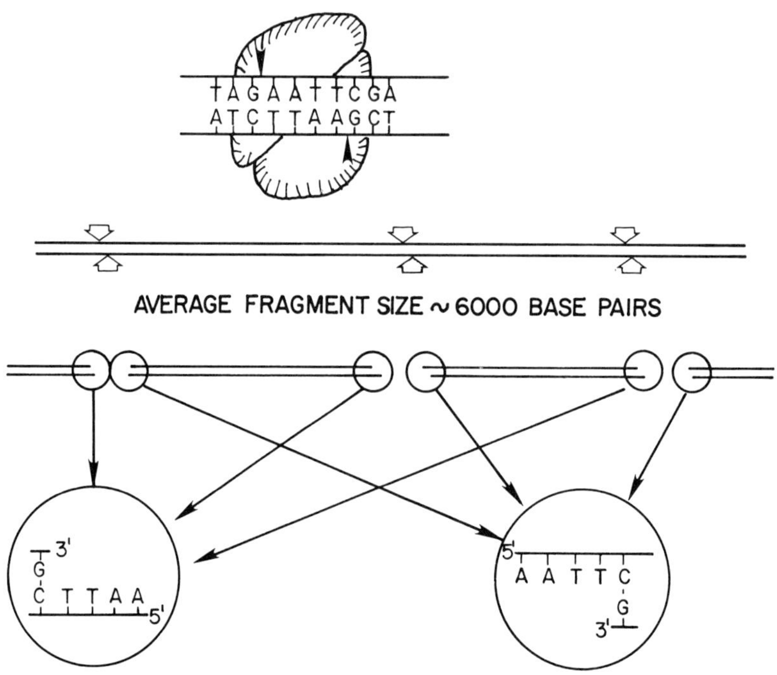


Figure 8. Cleavage of DNA restriction endonuclease $ECOR_1$.

base-pairing. Since all right-hand, respectively left-hand ends are identical, any DNA fragment generated by $EcoR_1$ can be joined by hydrogen bonding to any other fragment generated by the same enzyme. The link formed by hydrogen-bonding is not very stable, but can be transformed into a covalent bond by the enzyme DNA ligase (Figure 9).

The DNA fragment to be cloned and amplified is integrated into a so-called vector, either bacteriophage or plasmid DNA. Examples for plasmids are the colicin factors or the drug resistance factors, which have chain lengths of 6,000 to 30,000 or even more base pairs [as compared to the $3 \times 10^6$ base pairs of an E. coli cell] (Figure 10).

Plasmid DNA can be artificially introduced into E. coli, where it multiplies in parallel with the host cell DNA. Its presence in the host can be recognized by one of several genetic markers such a plasmid can carry, for example, resistance against an antibiotic such as tetracycline.

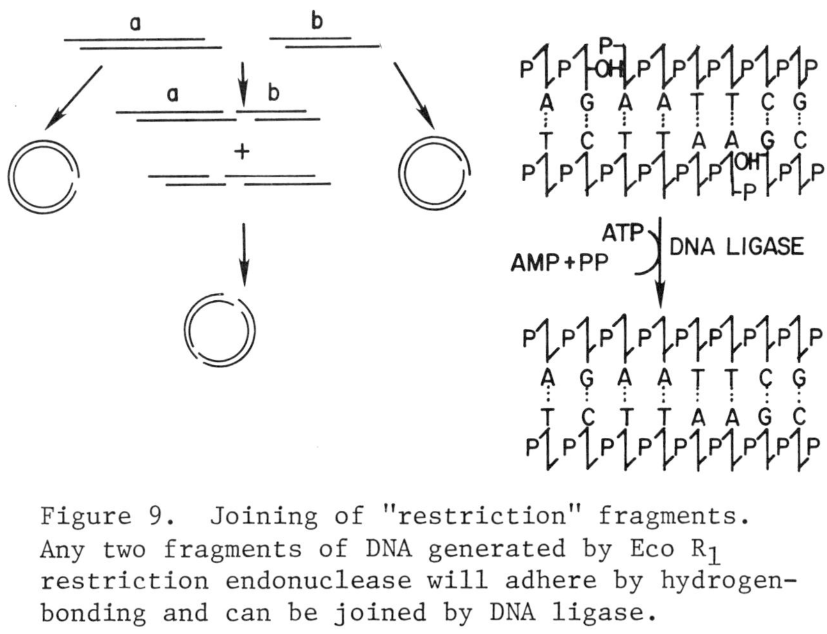


Figure 9. Joining of "restriction" fragments. Any two fragments of DNA generated by Eco $R_1$ restriction endonuclease will adhere by hydrogen-bonding and can be joined by DNA ligase.

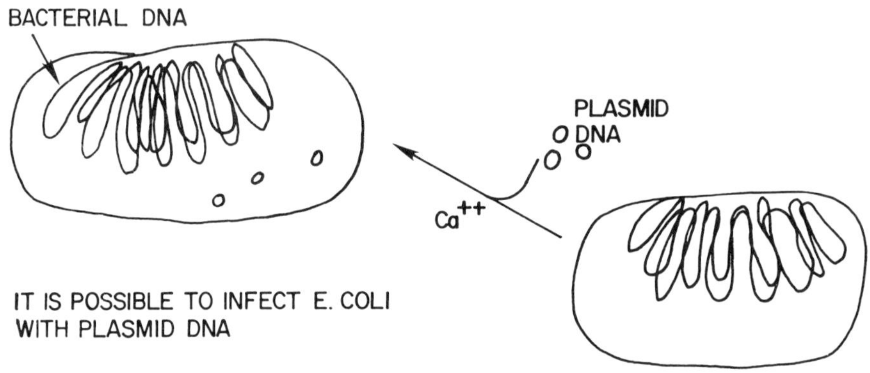


Figure 10. Plasmids of E. coli.
Independently replacing DNA molecule - may carry one or more genetic markers such as drug resistance. Molecular weights range from 4 to 20 x $10^6$ (coli genome 2 x $10^9$.

In an experiment carried out by Morrow *et al.* (Figure 11) DNA was extracted from a frog, *Xenopus*, and the DNA fraction containing the ribosomal genes was purified. A plasmid (R factor) DNA, containing the tetracycline resistance marker, was also purified. Both DNAs were cleaved with the restriction enzyme $EcoR_1$ and then joined with ligase to circles. These chimeric or hybrid DNAs were then inserted into E. coli.

Bacteria which had acquired tetracycline resistance were selected, cloned, grown to large quantities, and the hybrid plasmid was extracted. This hybrid plasmid was cleaved with $EcoR_1$ restriction enzyme to yield the ribosomal DNA moiety and the plasmid moiety. Not only does this provide a method for producing large quantities of pure eukaryotic genes, but it also allows the cloning of DNA, i.e., the almost unlimited amplification of one single molecule. Potentially, any single gene could be cloned and amplified by this procedure but it is usually necessary first to purify or at least enrich it to some extent.

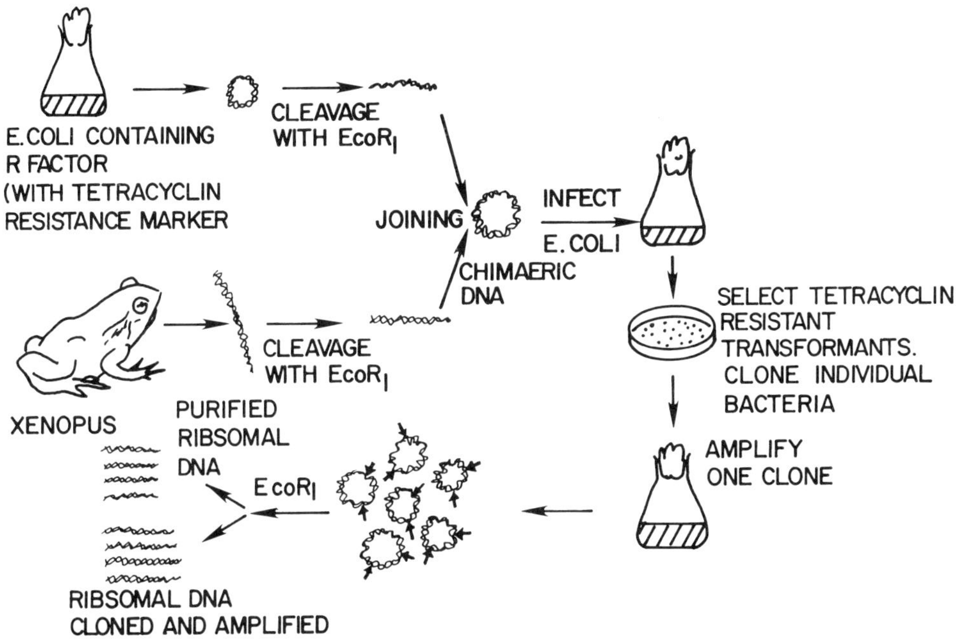


Figure 11. Cloning of an eukaryotic gene.

## Methods of Purifying and Enriching DNA

Several methods are available for this purpose or are being currently developed.

1) Physical separation procedures are based on the finding that certain genes, in some organisms, have a nucleotide composition which differs to some extent from that of the overall DNA. It is possible to enrich or purify DNA such as 5 S, histone, or ribosomal RNA-specific DNA by centrifugation in a CsCl density gradient, most advantageously in the presence of actinomycin, or $Ag^{+}$ or $Hg^{++}$ ions.

2) A scheme successfully pursued in several laboratories, including our own, starts out with the isolation of a specific mRNA. A DNA complementary to the mRNA can be synthesized with reverse transcriptase, i.e., RNA-directed DNA polymerase.

This single-stranded DNA can be converted into a double-strand with DNA polymerase and inserted into a plasmid. The complete gene corresponding to globin of rabbit has been prepared and its sequence determined.

3) The approach pioneered by Khorana is that of chemically synthesizing blocks of oligonucleo tides, which are then assembled into longer sequences with the help of DNA ligase. The synthetic gene for $tRNA^{tyr}$ has been cloned, as has the lac operator. To permit the efficient translation of eukaryotic DNA in prokaryotic systems, it may well prove necessary to combine the isolation of natural DNA segments with synthetic modifications.

4) In cases where the purification of a specific DNA segment is not easily feasible by physico-chemical or biochemical means the so-called "shot gun" approach has been used. The total DNA of an organism is fragmented by a restriction enzyme and integrated at random into plasmids. These are transfected into bacteria and a large number of clones are obtained. The DNA from these single clones is tested by hybridization with the labeled mRNA corresponding to the desired gene. With a methodical testing program large numbers of clones can be screened. This method has been used successfully to isolate histone genes.

## Potential Applications of DNA Cloning Techniques

The following cloning techniques may be employed:

1) physical and chemical studies on genes and the DNA surrounding them

2) investigation of *in vitro* expression of single genes and reconstituted chromatin

3) genetic mapping

4) reintegration of DNA segments into homologous and heterologous cells

- to study expression and regulation
- to generate microorganisms capable of producing valuable eukaryotic proteins (for example, hormones)

- to create improved crop plants (for example, containing genes for nitrogen fixation)
- to correct certain genetic deficiencies.

# *The Extraordinary Adaptability of the Human Fetus*

Pierre Royer

I wish, at the conclusion of this part of our discussions, to comment on some of the difficulties that we have encountered.

The term "adaptation" has different meanings among the disciplines represented and in the different languages utilized in the course of this Colloquium. These variations lead to extended discussion but they also weaken comprehension. The important thing, always, is that we be aware of it.

The statements we have heard had the ambition or goal of linking genetics with the physiology of regulations and finally with psychosocial processes. However, there is evidence that the methods of approach and the content of thoughtful inquiries are very distinct from each other in these different domains. We must accept this difficulty, even if certain arguments are by nature more by analogy than by basic principles.

To complement the statements, I wish to comment on the systems of adaptation and the capacities of adaptability that are expressed during prenatal and perinatal development of a human being. This very schematic commentary may provide a useful subject for reflection for our colleagues who themselves are not specialists in the field of biological development of man. They concern the extraordinary adaptability of the human fetus to drastic changes in exposure to food and oxygen which are indispensable to his growth and maintenance. The egg-milieu system undergoes four successive steps:

1. After impregnation, the egg multiplies, while traveling in the Fallopian tube, before being implanted in the maternal uterus. This period of preimplantation lasts about a week in humans. It is still called a phase of segmentation because the egg undergoes cellular divisions, bringing it to the stage of 32-64 cells. This cellular mass is free and mobile, but requires very little nourishment (secretion of the tube and of the uterine mucous) and oxygen. This cellular cycle is very special and cellular division is made almost without cellular growth, so that the mass of the 32-64 cells is almost identical to that of the egg before the first division. The adaptation of the cellular cycle to nutritional conditions is shown by the very delicate method by which a complete and adequate "cellular nutritional cycle" is probably selected because it is favorable to the movement.

2. At the end of the first week, the egg is implanted in the uterine mucosa and constitutes a maternal embryo allograft posing immunological questions not yet resolved. At this time of placentation, adaptations exist permitting exchanges of nourishment (water, electrolytes, amino acids, glucose, and oxygen) with the mother; for three months these exchanges are of moderate efficacy. During this period, adequate development of the embryo is mainly through growth by cellular multiplication and relatively little by cellular growth.

3. At the end of the fourth month of gestation, the placenta takes shape and expands. It permits a continuous perfusion of oxygen and nourishment to the fetus from the mother. The growth of the fetus is now very rapid; it is still by cellular multiplication but also by cellular growth. In addition, in humans, there is an excess of nutritional material permitting even the development of glycogenic and lipid reserves, and preparing the fetus for independent life.

4. Actually, the last step is birth. The human passes from a liquid milieu to an air milieu, from a continuous supply of nutriments by perfusion to a discontinuous one and by digestive

alimentation, and oxygenation by the respiratory path and from temperature regulation by an excess of thermal units in the fetus to one involving a struggle against the excess cutaneous and respiratory thermal losses.

These four changes of milieu schematized here--solely from the point of view of the availability of nourishment--constitute some constraints overshadowing those compared with man living in physiologically extreme situations in the postnatal period, a situation where one appreciates more readily adaptability and adaptation. From these successive adaptations of the human fetus to the types of available nourishment, the role of pre-established genetic organization seems considerable.

I wish to emphasize by this example that the adaptations to variable environments are not limited to physical or geographic or ecologically different systems, but can arise within living beings in the course of their proper biological development.

# *Adaptive Functions at the Human Level*

Jacques Ruffié

Adaptation is a very general phenomenon, since it is observed in all living species. It may be defined as the fact that the living being possesses organs, functions, and behavior capable of responding to the exigencies of the environment in which it lives. These capabilities allow the organism to assure its own survival, reproduction, and line of descent; that is, they assure the permanence of the species (6). Any maladapted or unadaptable individual has little chance for survival. Thus adaptation appears today to be inseparable from the idea of life itself.

According to Albert Vandel (3):

> Adaptation is organization itself. Adaptation thus corresponds simply to a certain manner of viewing the organization of the living being in its functional relationships or in its relationships to its environment and the beings surrounding it. In this sense, one might say that the problem of adaptation refers to nothing other than the origin of organization seen in its functional aspect.

The problem of adaptation is in reality a spurious problem, linked to an anthromorphic and rationalized view of the world. However, especially at the stage of *Homo sapiens*, reason is the most routine method of grasping reality. The phenomenon of adaptation appears to us, at least among the higher vertebrates and particularly in man, to correspond to three processes which are different but can converge to yield the same effect. Thus there exists a sort of synergism among phenomena proceeding by different routes or in totally unrelated domains. These three types of processes may be defined as follows: 1) adaptation in its proper sense, or genetic adaptation; 2) acquired adaptation, also called acclimatization of habituation; and 3) sociocultural adaptation.

## Genetic Adaptation

This refers to a phenomenon which has been fixed by heredity and obeys a selective model. The resistance of bacteria to antibiotics, that of insects to pesticides, and numerous characteristics observed in higher vertebrates belong in this category. It is on the basis of genetic adaptation or maladaptation that the Neo-Darwinian theory of evolution is founded. Definitely fixed in the genome by unique information and perpetually transmissible, the adaptive character persists even when the selective pressures which formed it have disappeared. The wings of birds, the fins of fish, and many other biological characteristics are adaptations. Such adaptations permit the camel to survive in the desert and llama to live at high altitude (4).

## Acclimatization

This refers to an acquired response which follows a stimulus but is not genetically programmed. It disappears when the causes which elicited it disappear. It is thus a purely physiological and situational phenomenon. The classic example of this is the polycythemia of altitude. The subject arriving in the high mountains multiplies his red blood cells. This erythrocythemia disappears when he returns to sea level.

## Psychological Adaptation

The third type of adaptation consists of a conscious voluntary response to the exigencies of

## REFERENCES

1. Mayr, E.: _Populations, Espèces et Évolution_. Hermann, 1974.

2. Ruffié, J.: _De la Biologie à la Culture_. Flammarion, 1976.

3. Vandel, A.: _l'Homme et l'Évolution_. Gallimard, 1949.

4. Anthropologie des populations andines. Paris: _Colloque INSERM_, 1976.

5. _Biosocial Interrelations in Population Adaptation_. Mouton, 1976.

6. Préadaptation et adaptation génétique. Paris: _Colloque INSERM_, 1971.

# *Psychosociology of Adaptation: Role of a Natural Phenomenon (Color Phenotype) in the Process of Cultural Adaptation*

Francois H. M. Raveau

The term adaptation in this communication will designate all forms of interaction referable to the functioning of an organism, a personality or a group. We must also regard adaptation as a bilateral process developing in a system of permanent interaction wherein various regulators exist. The adaptive process will take place in an environment of continuous behavioral activity, rather than being a disruption in homeostatic equilibrium, be it biological or cultural. For this reason we shall confine our reflections to usual categories, such as the change in the stage of development of personality (in analytic terms, for example, the change from the anal to the genital stage). Neither shall we relate adaptive behaviors to certain categories which disrupt equilibrium (climate, affective frustrations, etc.); but we shall consider some causes of disruption of equilibrium, the action-reaction engendered, and consequent changes in the behaviors of the group under study (conduct, attitudes, reactional opinions, etc.).

The particular conditions which our study reveals occur only in the experience of migrant populations. In effect, they present a disruptive system which becomes a new environment. The observation of the crisis, induced by leaving a place in which the homeostatic biological and social processs are generally effective, to go to a new place in which the customs and standards are still to be learned and practiced, provides a quasiexperimental field of study.

For our purpose, among the different types of migration which our society suggests today, we have selected a group of Antillean workers, a sample, even with all of its ambiguities, which furnished our investigators an excellent field for study. It was possible, at the same time, to survey both internal and external migration since the sample consisted of French nationals (an institutional distance of zero), colored people (maximum phenotypic distance), and transplanted individuals (bioclimatic and dietetic distance).

Let us specify that we are dealing with the Antillean personnel of the Necker Hospital. We have surveyed this group by means of a questionnaire interview which has furnished us with information covering both the course of adaptation and its implications. The typologies of the adaptation will consist of the etiology and semiology of the migratory adventure which defines a mode of adaptation to a specific environment consisting of reactional subenvironments. The indicators, furnished by the 156 multiple choice questions and 23 open questions, cover both explanable variables (furnished by descriptive elements) and unexplained variables (behavior, attitudes, and opinions).

It is thus that the conditions of migration: housing, study-work, budgets, health, civil status, language, leisure, social relationships, religion, political interests and participation, sexuality, and perception of adaptation, were obtained. More specific data, suited to the Antillean environment and innate to this ethnicity, were added, e.g., the phenotype, relation to territory (status, etc.) and language (Creole).

The sample comprised 321 subjects. We shall present a global rather than a detailed report of the data by visualizing the structure of the collection of variables derived from the survey. To do this we shall proceed to an analysis of the responses (to the questionnaire), the object of which is to represent a profile of individuals (or of variables) in a rather limited space, while maintaining a maximum of information. A small number of variables, called factors, have been selected and calculated for each individual and each variable. The indicators, called contributions, permit the measurement of the relative part of each of the variables in the determination of these factors and are, thus, directed at the interpretation of an axis. We must once more point out that these factors are not correlated and hence express independent tendencies.

Four figures of graphic nature revealing those diverse variables, which have been significantly retained, are presented. We then conclude with a global diagram which shows the interdependence of the various factors.

## Ethnicity

Figure 1 shows the indicators representing different channels which permitted the examination of a number of individuals. The first evaluation is of the phenotype. It is quite evident that the social review of a phenomenon of biological nature, such as the melanoderma, is complex. In our effort to understand how the color could be transformed into cultural facts, we must not overlook the difficulties inherent in the relocation of our subjects.

Three evaluations of the somatic color and the social color of our population were combined with the question covering adaptation. The first of these was done by investigators who divided the 321 subjects into two groups: one with light skin and one with dark skin. The two others were those done by the subjects themselves in response to the question, "Could you tell to which type you belong?" They defined the type to which they belonged in their original culture. These categories emerged as specific terms for an Antillean typology of appearance, most often based upon the color of the skin, but complicated with morphological traits and, by progressive shifting, with sociocultural elements. The etymology of these terms must be sought in the recent history of

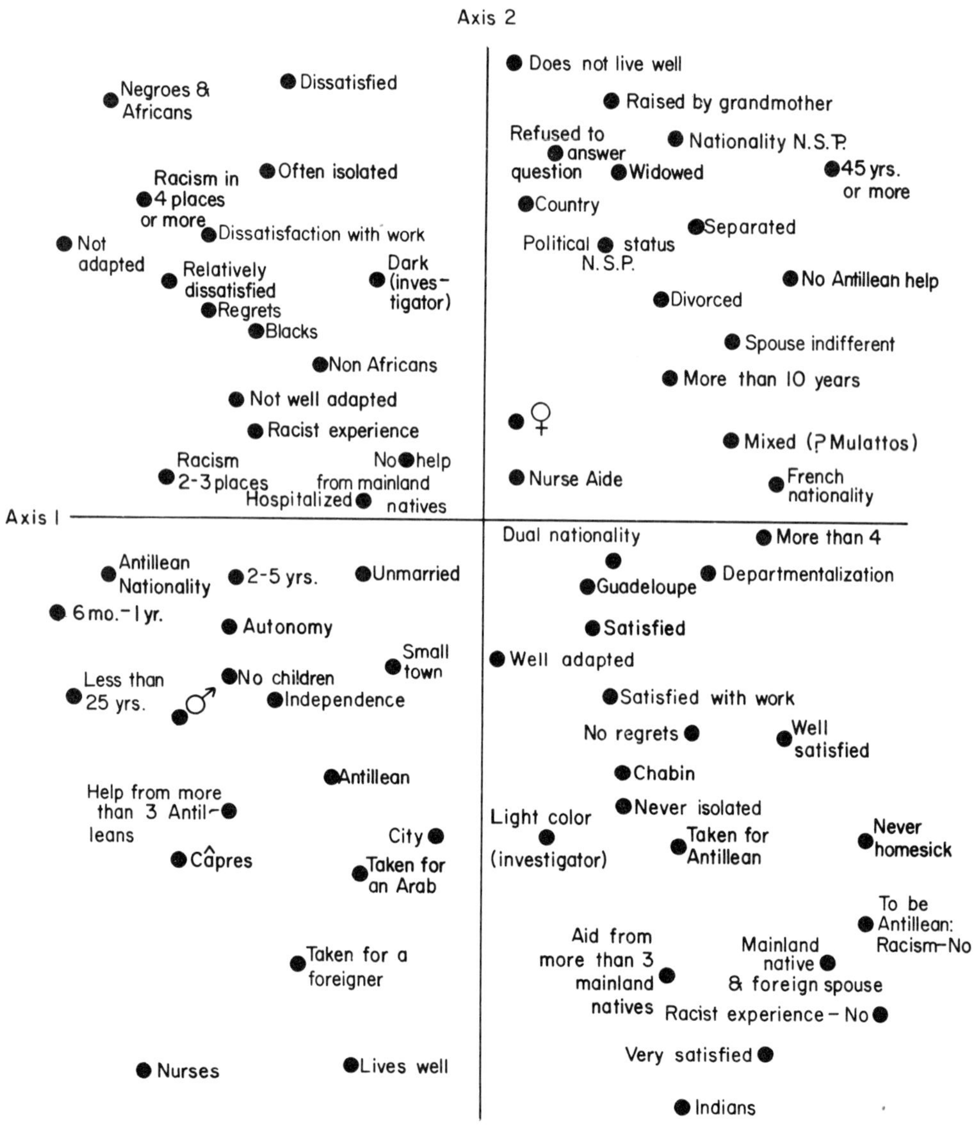


Figure 1. Ethnicity.

trade or of slavery. Complex relationships have been established in a society in which cross-breeding with Africans and Europeans has progressively modified the original genetic strain. In responding to the question, "When a French native does not know you, for whom does he take you?" they indicated the identity which they believed that a mainland native would make in response to their appearance. This was a mirror through which the Antillean saw himself in the eyes of the French people. The response given reflected not only how the subject felt he was perceived by his host society, but it also indicated the image he would choose to reflect. The migrant first identified himself in terms of his original culture and then in terms of his life in the city. Reference to this double identity appeared to be constant in the migrant's self-image. It could be humiliating if he were seeking an identification with French society; or it could be transcendent if he were seeking to detach himself from it.

These three forms of evaluation evoke a certain number of comments and some reservations. First, from the information gathered in the course of the survey, an attempt was made at homogenizing the judgments of the three investigators both during the presurvey and during the group's work. The dichotomy, light/dark, was obtained by taking only the color of the skin into account. The facial morphology and the appearance of the hair were not considered. It was thought esthetic interventions such as straightening the hair and cosmetics, would confuse the charts since a large part of our population was young and female. The classical evaluations of the eversion of the lower lip, the width of the nose, the shape of the eye, etc., were perceived as leading to too many complex and inaccurate associations, and, hence, to errors. We did not employ the Vallois or Tisserand-Perrier scales since such an approach would have been considered as reified by our population, resulting in resistances which could have interferred with the progress of the survey.

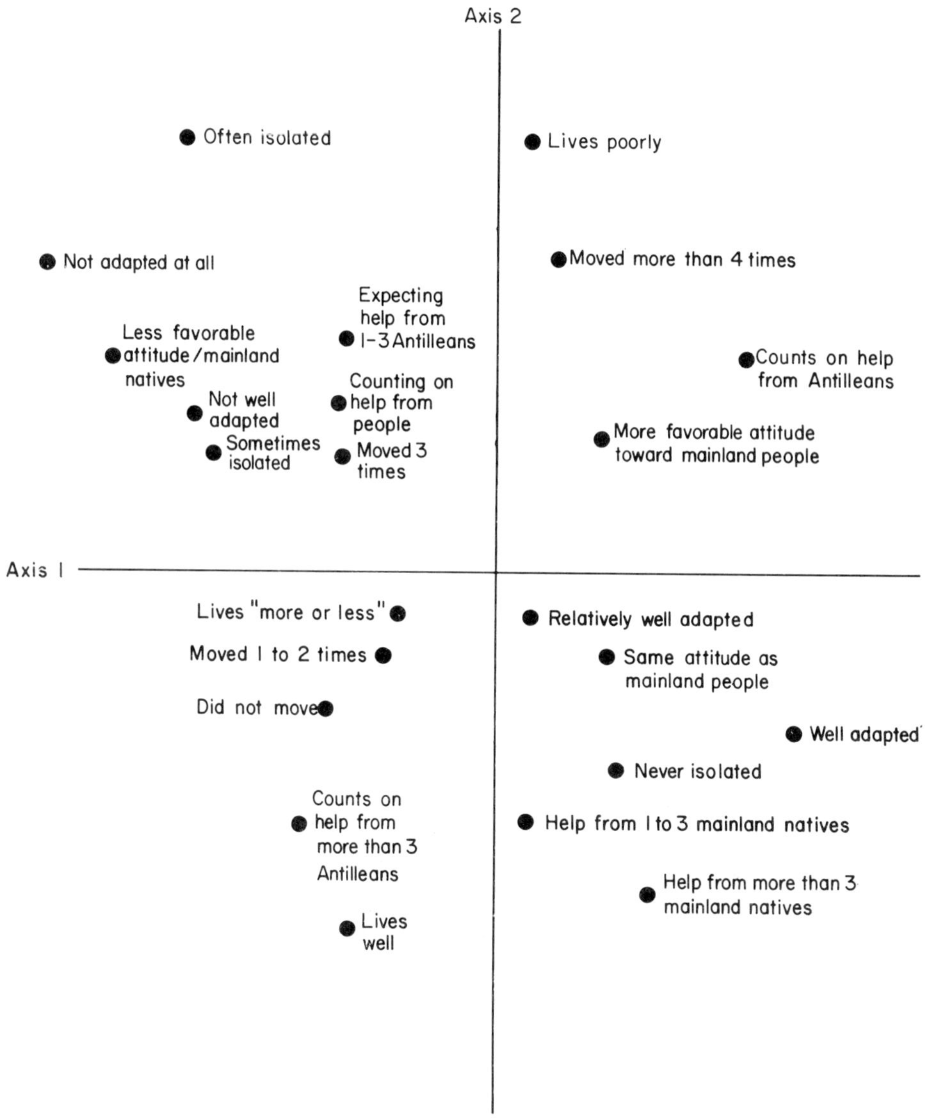


Figure 2. Adaptation.

## Adaptation

Figure 2 synthesizes various indices accepted by the subjects and recognizes the geographic mobility, isolation, relationship to help, and opinions regarding the city environment. The choice of spouse and of neighborhood appear to determine the degree of loyality to the group.

## Civil Status

In Figure 3 the usual indicators of civil status appear: age, sex, duration of sojourn, marriage, children, educational level, occupation, and income.

## Satisfaction

Various ways of perceiving the environment are isolated in Figure 4: length of stay, work housing, etc.

Finally, we have projected the global indicators in Figure 5. We shall comment on the first two factors, which are also expressed in Figure 1. Here only a few modalities of response are presented; these contribute most to factors one and two. On the right hand side of Axis 1 are seen the indicators of a good adaptation, the absence of racist experiences, and a certain identification with France (choice of departmentalization and of the French nationality in case of Antillean independence). The <u>light</u> subjects are found in this category.

Opposite to Axis 1, the most important variables are the choice of the Antillean nationality (Guadeloupe or Martinique), the desire of autonomy for the islands, the recognition of the difference with respect to French nationals, and the dissatisfaction of living in France. On this side of the axis the subjects described as <u>dark</u> are found.

On Axis 2 in the upper part of the figure are seen the variables indicating racist experiences and poor socioeconomic conditions. The <u>dark</u> subjects are found in this quadrant. In the lower

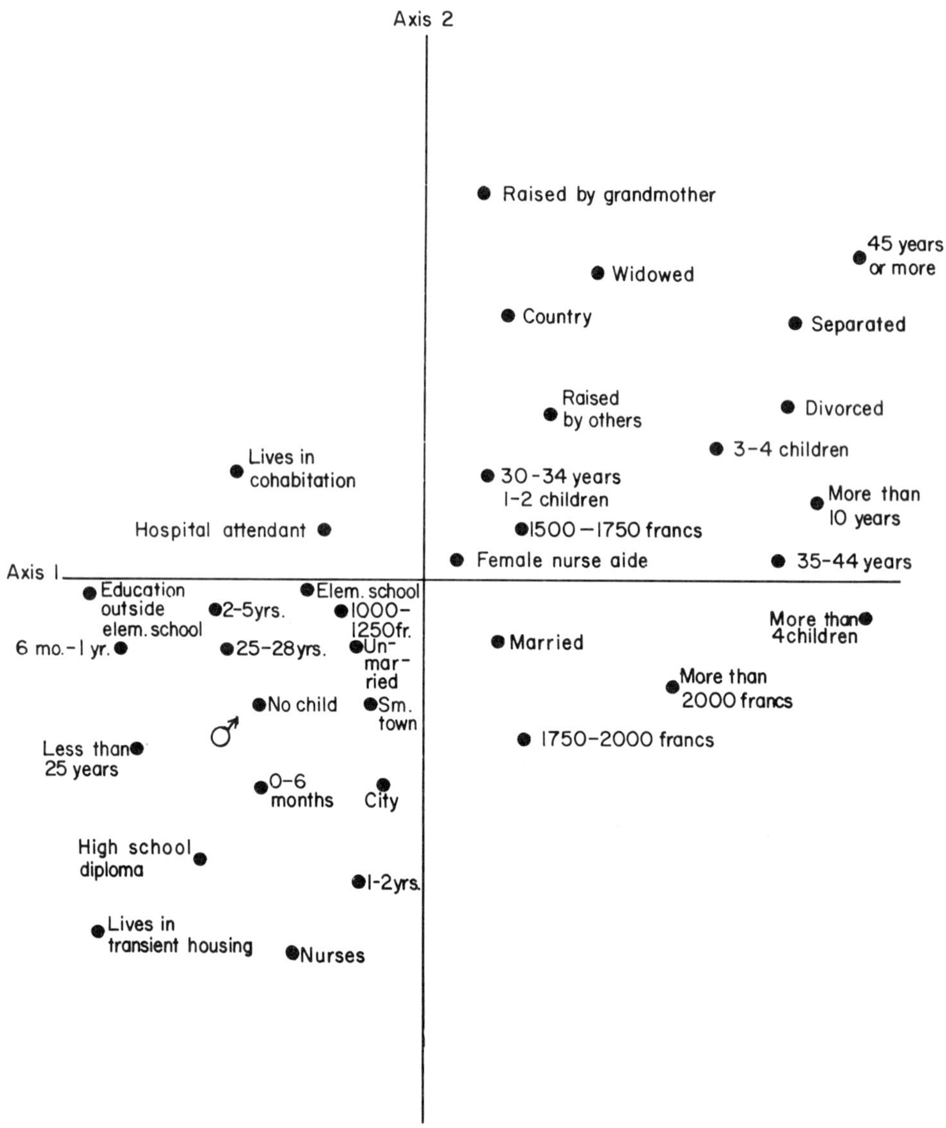


Figure 3. Civil status.

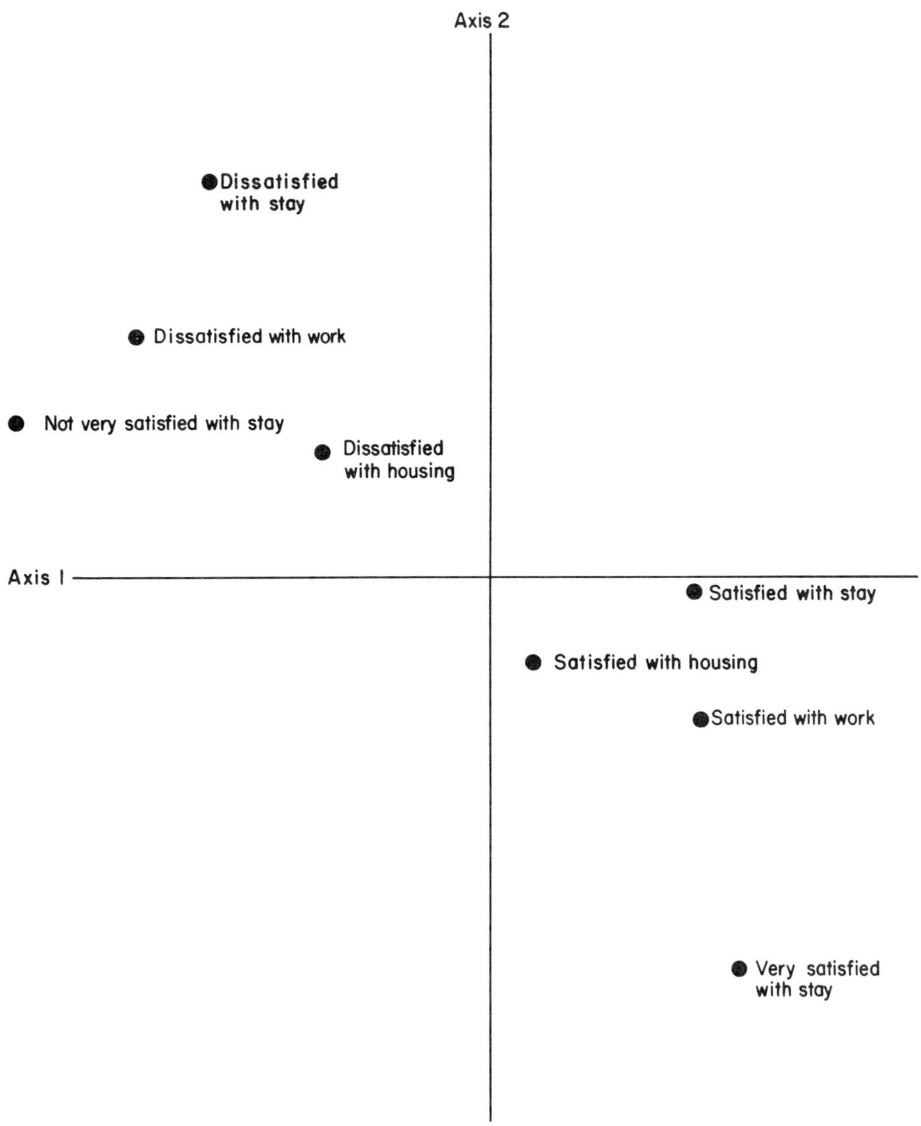


Figure 4. Satisfaction.

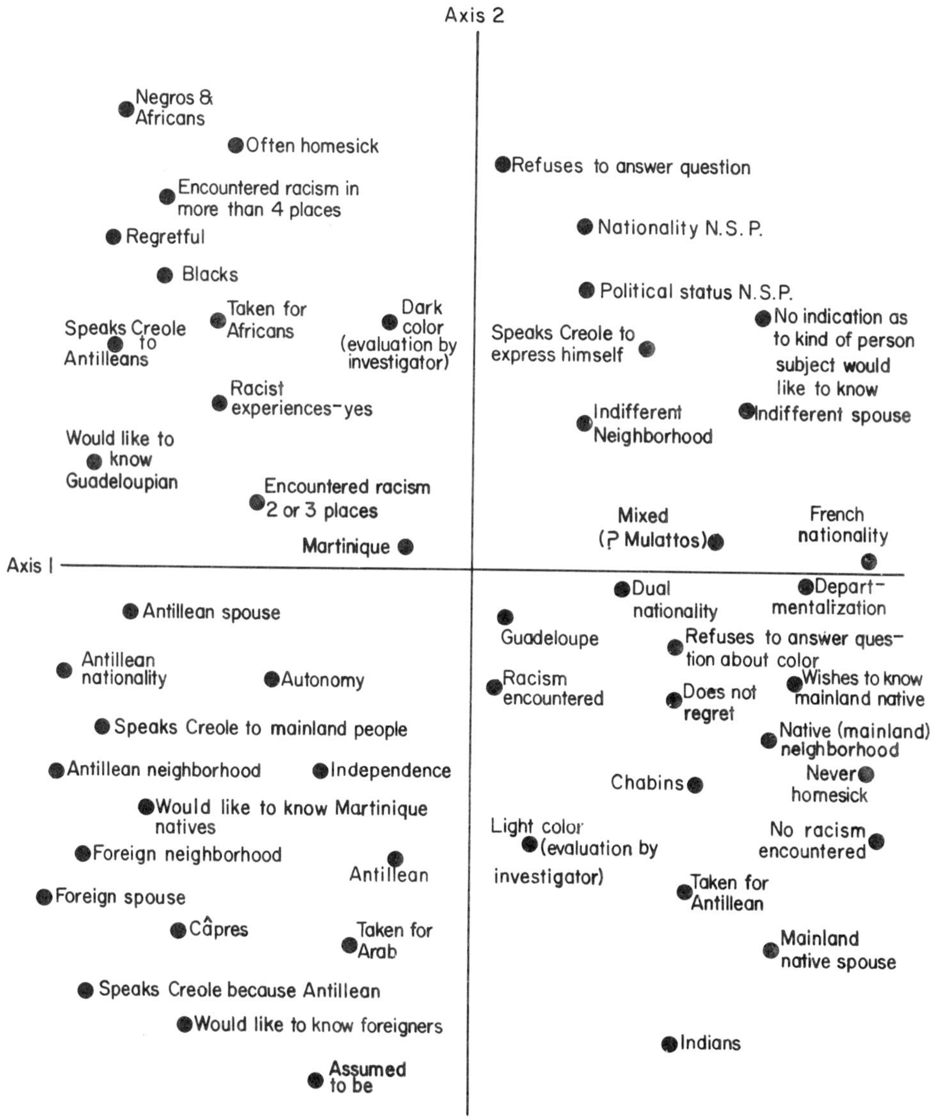


Figure 5. Global indicators.

half, where living conditions are considered satisfactory, the light subjects are found. Thus, it appears that skin color is associated with a whole series of indicators in the adaptive process: positive when the skin is light and negative when the skin is dark.

In the upper left quadrant, which indicates dissatisfaction and maladaptation, are found Africans and other Blacks, as well as subjects who stated that they are taken for Africans. In the opposite quadrant, where satisfaction is indicated, Chabins and Indians, as well as those who claim to be taken for Antilleans, are found.

However, satisfaction linked to a good life during the sojourn and, to a certain extent, to criteria of adaptation, is not wholly attributed to identification with the white race but, rather, to the acceptance of the migrant by his white host people. It is actually between these two quadrants, regrouping respectively the dark and the light subjects, that the culture-color dialogue is most manifest, the other (variables) being representative of lesser nuances (or differences). It would appear, in effect, that relatively favorable circumstances surrounded our migrants' lives during their sojourn in France in spite of their color which, in this study, has appeared as a handicap.

These favorable circumstances seem related to the lack of desire on the part of the migrants to identify with the French and their (the migrants') search for independence. This is shown in the lower left quarter where the inhabitants of Capres and Antilles are found. On the other hand, the absence of identification and its repercussions are illustrated by the upper right hand quadrant which relates a poor life to a difficult identification. It is in this quadrant also that one finds the subjects who refused to respond to the question, "When the French native does not know you, for whom does he take you?"

Figure 5 stresses the fact that the role of color is foremost in the dialogue of migrants with the host society but that defences against a humiliating subjugation are possible. The *light* subjects are less victimized by racist experiences within their French environment and are more inclined to acknowledge that the fact of being Antillean does not necessarily provoke racism. They declared themselves in favor of a law which would accord greater autonomy to the islands. The *light* subjects are better educated; a large number of them hold primary or secondary school diplomas. They were the developers of Antillean cities and have lived in urban communities longer. Finally, a larger majority of them asserted that they feel they are not different from the French natives.

The *dark* subjects seem to present the opposite picture. They have experienced more racist treatment from the French and more of them felt that being Antillean provokes racist behaviors. A larger number of them did not express concern about the status of the islands. Slightly less than half of their population have received diplomas, and they originated in rural districts. They represent a more recent migratory wave. Their families are definitely among the most disadvantaged, economically. They perceived themselves as being more different from the French natives than did the light colored persons. Here color has a double significance: it reflects a state of being Antillean and explains the process of adaptation in France. Those of the lightest color are educated, more urbanized, and are descended from familial environments of higher resources. Their independent social etiology explains the quality of their integration to the French community. To this is added their longer residence in Paris. They have probably profited from the lower saturation of the Parisian environment by the Antilleans because of their being less "visible" (lighter, therefore closer to the French). This reduction of phenotypic distance may also explain their easier identification with the environment: differing very little from the French natives, they are less the object of racist aggression.

In conclusion we should point out the inadequacy of the adaptation and satisfaction indicators because it appears that the psychosociologic exaction of a successful adapation must be paid for by less satisfaction, a clear statement of the effort demanded. We should also note the stability of the intermediate group in these classifications: the Antilleans are at the same time the most satisfied and claim to be better adapted than the Africans, a group which appears to be the most rejected by the host culture. This may be considered a real psychological cost of adaptation. Its evaluation may be obtained by taking to account the phenotypes in their relation to the indicators of satisfaction and adaptation.

The social evaluation of a natural phenomenon, such as the meanodermic phenotype, permits us to understand how color is actually transformed into a cultural fact. The relative darkness of migrants, as perceived by the host society, will determine the quality of the course of adaptation, but a relationship of reciprocity will be established between the one seen and the seer. This inspection, in its turn transforming the person being observed, can either crush him or cause him to become independent.

This study represents an incomplete account of the complementary relationships which are established when two cultures, entering into contact, promote adaptive interactions wherein the phenotypic (natural phenomenon) distance is eliminated by the cultures involved.

# *Genetic Adaptation and Preadaptation*

Georges Larrouy

The terms adaptation and, in particular, preadaptation are accepted and used to denote a variety of meanings, often inadequate ones. In this brief report we will first offer some definitions accepted (but often misunderstood) by physiologists and specialists in ecological genetics. Having thus established a precise basis for discussion, we shall attempt to outline very briefly the general evolutionary significance of adaptation and preadaptation. Finally, we shall attempt to evaluate the importance of these phenomena at the level of the human species, using some well known and instructive examples.

The adaptation of an organism to a particular environment is manifested in its ability to develop and reproduce itself in this environment. The word adaptation, when used in the singular, denotes a multifaceted phenomenon carried to a more or less successful conclusion. However, when the same word is used in the plural or preceded by an indefinite article, its meaning is far more restrictive (Molinier and Vignes). Adaptation requires genetic preadaptation; that is, the presence in the genome of certain genes capable of determining or defining an individual whose physiology will be in harmony with the surrounding environment. In the phenomenon of preadaptation it can be seen that pure chance plays a significant role by continually altering the genes, while necessity governs the ensuing selection of genomes.

Adaptation, the perceptible result of environmentally determined sorting of the individual genomes that constitute the preadaptive variants, is sometimes difficult to distinguish from *post-adaptation* or *accommodation*.

The latter terms refer to the degree of freedom of phenotypic response (molecular, physiological, or morphological) in relation to genetic information, under more or less exceptional conditions.

Despite the possibility that an organism may achieve a more or less exact accommodation to the environment with respect to particular functions or metabolic compartments, the survival of a species is seriously threatened by brutal modification of any of the parameters defining and delimiting its ecological niche within its geographic range. Extinction is most likely to occur when the niche is highly specialized or the range severely restricted, because these conditions permit a subtle or localized environmental change to exert an effect on the entire population of the species.

Therefore, in order to have a reasonable chance for long-term survival and some prospect of an evolutionary future, a species must occupy and utilize the maximal variety of ecological niches (perhaps transforming them in the process) within the largest possible geographic range. Then, given abundant and readily utilizable trophic resources and available space, the species will increase its effective population size together with its possibilities for genetic recombination. Thus it will have acquired a source of new preadaptive variants and ecological plasticity. In this context, we should note that a vast geographic range and occupation of varied ecological niches must have arisen, in the case of a young species, from a certain multivalence of function (e.g., omnivorousness). When there is no motive force, environmental or ethological, to cause range disjunction, genetic mixture will once again favor permanent diversification of genomes. This diversification, or genetic heterogeneity, is a factor contributing to the success of the species. As long as gene mixture is maintained at the level of large and cohesive populations occupying varied environments, adaptive specialization is limited to a few individuals and manifests itself only as a chance effect. But when significant modification of the environment occurs, diversification of genomes allows the species to respond appropriately. The response consists of differential survival and selective reproduction of individuals having physiological capacities fitting them to the new environment. Conquest of new territories occurs through the same process.

Preadapted and selected individuals, although better suited to the new demands of the environment, are more specialized and more limited in their functional capabilities than the parent species. The new population is more monomorphic and its variability is lower (Mayr). If the environmental changes persist, the result will be a rapid ethological and then genetic segregation which will give rise to a small species branching off from the parent stock. The maintenance of a given selective pressure, acting in a precise manner on a stock of genomes capable of furnishing suitably adapted individuals, will eventually result in an impoverishing speciation--a type of specialization which can lead to evolutionary obsolescence.

Man is a species whose past and present history can be used to illustrate the points made in this introduction.

The nutritional plasticity of the human species has allowed it to utilize a great variety of resources. If one adds to this plasticity a technology whose growth permitted the remodelling of increasingly populated environments which might bear little resemblance to the ideal niche, together with the elimination of competing species, then one understands that the range of the human species has increased rapidly along with the size of its population and, therefore, the variety of individuals under the influence of genetic recombination.

Nevertheless, within the human species, there have appeared certain distinctive and genetically coherent groups with more or less specialized features. These groups have arisen as a result of the sorting effect of certain factors that are characteristic of the highly diversified environments populated by the human species acting upon gene pools presenting the appropriate variants. Attempts at typological classification of "races," together with elaboration of some already old-fashioned laws (law of sizes and Bergmann-Allen's, Gloger's law, etc.), whose applications to the human species have been numerous and not always baseless, reflect the phenotypic expression of some adaptive coherence. There are still some selective factors of the environment from which neither technology nor culture permit man to escape. Among these we may mention parasitoses, on account of their high frequency, indeed virtual omnipresence, in those geographical zones within which primitive mankind developed and diversified. As a second example we may take high altitude, an environment affecting millions of humans over a great many generations.

In many regions, especially the tropics, parasites in the broad sense (viruses, bacteria, multicellular parasites, fungi, etc.) represent one of the most active and selective environmental parameters acting on animal species in general and humans in particular. Parasites exert a very strong selective pressure on individual genomes. Jacques Ruffie and I have thought for some years that certain genetic traits of human races are the result of selection by parasites of traits which conferred some advantage in a given context. Many specialists now incline toward similar hypotheses, which may be supported by any number of examples.

The foremost example is undoubtedly the case of abnormal hemoglobins. Certain hemoglobinopathies have a distinctive geographic or racial distribution. Moreover, in certain regions, these genes occur in frequencies considerably in excess of those which would be predicted if chance alone were the determining factor. Allison and other Anglo-Saxon authors offered the hypothesis that these abnormal hemoglobins might confer protection against some selective factor in the environment.

It appears that, in the zones in question, the selective factor is undoubtedly malaria, particularly that caused by *Plasmodium falciparum*. In the case of hemoglobin S, for example, homozygotes (SS) are destined for premature death because of their severe clinical difficulties. Homozygotes possessing a normal hemoglobin of type A are susceptible to attacks of malaria and frequently die during childhood. However, heterozygotes (AS) have much less severe attacks and can therefore survive to reproductive age. Thus heterozygotes have the best chance of reproducing and are likely to transmit the sickle-cell trait to their progeny.

Similarly, there are certain populations (the Mnong, Thai, Khmer, etc.) of the Indo-Chinese peninsula that have the gene for hemoglobin E in extremely high frequencies. These peoples have long inhabited regions infested with *Plasmodium falciparum* malaria, although they have been displaced from other areas by peoples of more northern origin. Until now, the ethno-political map of Southeast Asia has been a concrete expression of certain epidemiological realities.

However, as noted before, genetic adaptation has its price: Although AS heterozygotes are favored over AA homozygotes in malarious environments, they are at a disadvantage elsewhere. Work at even moderately high altitude or air travel may induce severe sickling attacks in bearers of the hemoglobin S trait.

Epidemiological information now allows us to explain the peculiar distribution of a number of genetic traits which are, in a protected environment, more handicapping than favorable. These include glucose 6-phosphate dehydrogenase deficiency and certain electrophoretically defined serotypes such as HpO.

These mutations, although generally considered as handicapping or neutral, are relatively favorable in certain environments, in which they may be selectively transmitted in a genuinely preadaptive fashion. However, in our medically and technologically advanced society, these traits represent grave handicaps and may acquire a pejorative connotation. Enzymatic deficiencies likely to cause serious medical complications clearly illustrate the relativity of the selective value of a mutation.

Epidemiologic context determines the distribution of many other genetic traits having preadaptive value with regard to a particular environment. Ruffié and I proposed such a working hypothesis to explain the fixation of blood type O among South American Indians.

Recently, evidence has been presented which suggests a possible protective value of the Fy gene against Plasmodium vivax malaria.

There are now many reports suggesting interrelationships between specific diseases and blood antigens such as ABO and HL-A. At times when a disease such as smallpox has exerted a strong selective pressure on a population, it is highly probable that one or another genotype has acquired a very real preadaptive value.

In the case of balanced polymorphism, as represented by some blood markers, the situation is somewhat conjectural. An adaptive mechanism of this sort involving only one locus represents an exceptional case. If we have many such examples, it is only because they are easier to characterize than are balanced polymorphisms involving two or more loci. However, it is likely that adaptations involving polygenes are far more common, although these are difficult to detect and their selective value is less easily demonstrated. At first it may be difficult to estimate the role of postadaptation or accommodation in such situations. However, after a population has evolved for generations in an environment exerting a "stable" selective pressure, it will become clear that a significant number of individuals show adaptive modifications. The extent and diversity of adaptive changes will of course depend upon the genetic "keyboard"--the individual genomes which are acted upon by selection. Accurate analysis will indicate that the mechanisms involved in adaptations may differ, with selection occurring most often through the summation of slight modifications and/or adjustments, which may be morphological, physiological, molecular, or even cultural in nature. Ultimately, the adaptive result may be equivalent in spite of the diversity of individual solutions. Multiple enzymatic heterozygosity seems to play an important role in genetic adaptation, ensuring flexibility of metabolic response.

In the case of life at high altitude, the radical and permanent reduction of atmospheric pressure relative to that which exists at sea level represents an environmental parameter from which technology offers no lasting protection. It is natural that populations subjected to this transition should exhibit evidence of it, in the form of adaptive changes at various physiological levels. The imprint of adaptation is quite different from the signs of accommodation or acclimatization observed in subjects briefly exposed to high altitude.

Certain morphological or macrophysiological modifications occurring in populations living at high altitude may be attributable in part to accommodation. These include changes in the diameter of the thorax and the length of the sternum, elevation of blood pressure in the pulmonary artery, and thickening of the muscular layer in the arterial wall.

However, it appears that certain biochemical modifications may represent genuine adaptive changes that are of genetic origin and considerable selective value. For instance, the increase in activity of certain glycolytic enzymes that catalyze the synthesis of 2-3 DPG and promote $O_2$ desaturation of the blood at the tissue level appears to be an adaptive modification. These glycolytic enzymes are balanced by an array of diaphorases which permit maintenance or utilization in the blood of a significant quantity of methemoglobin, the stability of which is well known; but individuals bearing this specializing adaptation pay a price for it, in the form of a reduction in flexibility and a tendency to respond adversely to rapid changes in atmospheric pressure. Their great sensitivity to reduction of partial pressure of $O_2$ may cause them to suffer acute vascular accidents when undergoing a rapid transition to a higher altitude. These accidents are due to drastic elevation of blood pressure in the pulmonary arteries above the alveoli.

A precisely and quantitatively definable environmental influence affecting an entire population, such as the reduction of atmospheric pressure at high altitude, represents an unusually simple, clear-cut case of a selective factor. It seems reasonable to assume that adaptive solutions achieved under the influence of more complex, more variable, or more localized selective factors would be highly diverse and almost impossible to characterize. Moreover, we cannot easily detect or evaluate the role of factors like gene penetrance and expressivity, pleiotropy, and epigenetic phenomena in any given adaptation.

An adaptation is a balance, but the mechanisms giving rise to this result depend on the preexisting array of available genomes and thus upon preadaptive factors, together with environmental selective factors, which are often elusive in their diversity and variability. These various factors differ in their potential for and rate of change, so that the adaptive outcome is only a short-term solution which more or less successfully approximates the ideal solution. The individual (of the genome of the individual) is always the actual target of selection. The genetic structure of the population is the temporary result of selection and we can perceive only "snapshots" of the selection process--glimpses which make it difficult to reconstruct the entire "motion picture" of evolution, with its constant variations involving complex factors.

In conclusion, it can be said that man, when he has been unable to escape major environmental pressures through his technological resources, has responded like every young and demographically expanding species, by producing individuals with appropriate genetic potentialities selected by the environment. Specializing adaptation has thus occurred in the course of human evolution. This phenomenon is manifested in the homogeneity as well as the morphological and physiological distinctiveness of certain human groups. The persistence of specializing adaptation aggravated by some degree of cultural segregation, might eventually have led to evolutionary obsolescence for these groups.

However, the increasingly wide distribution of technological and scientific progress, together with the amplification of genetic admixture now underway, will restore the evolutionary youthfulness of the human species, help it to maintain its polymorphism and variability, and enable it to escape for some time from specializing adaptation and evolutionary obsolescence.

# Index